PRAISE FOR *LEADERSHIP IN THE REAL WORLD*

"A leadership read that's based on real-world experiences AND is actually enjoyable!"

— John Schmidt, Colonel, USAF (Retired)

"I was on my third Air Force assignment as a Forward Air Controller (FAC), and Tom was on his second; it was a steep learning curve for both of us, and we know the leadership and decision-making skills demanded by FACs in combat. They had to tightly control the air assets while being bombarded by calls on three different radios. FAC decisions had to be quick and accurate — a wrong decision meant fratricide or escape of the enemy. Tom brought that unique decision making skill to each of his future leadership situations."

— Jim Palmer, Colonel, USAF (Retired)
Vietnam FAC 1968-1969

LEADERSHIP
IN THE REAL WORLD

TEAM WORK
PEOPLE
DECISION MAKING
STANDARDS
MISSION
LOYALTY
DELEGATION
COMPETENCE
COMMUNICATION
ETHICS

50 YEARS OF BUILDING A LEADERSHIP MODEL

BY TOM PETITMERMET, COLONEL, USAF, (RET)

TACTICAL 16
PUBLISHING

Leadership in the Real World:
50 Years of Building a Leadership Model

First Edition

Published by Tactical 16, LLC
Monument, CO

ISBN: 978-1-943226-57-3 (paperback)

CONTENTS

FOREWORD

Leadership has many different yet similar definitions. For the purpose of this book, leadership is the effort given by either the officially designated leader of an organization or one of many informal leaders who sometimes emerge in any organization to accomplish an assigned mission or task. The exact genesis of the leadership tenets in this book come from a very successful twenty-six year flying career in the U.S. Air Force, fifteen years as a successful defense industry consultant, and more than twenty-five years as the leader of a nonprofit volunteer group. In this book you will find real-world examples of leadership in various situations that you can apply in your leadership situations. These fifty-plus years of actual hands-on leadership situations formed my leadership philosophy. I decided to write this book about leadership as I have observed so many recent instances where both our military and civilian leaders have taken paths that are completely opposite from the leadership tenets that I know and have practiced for the past fifty-plus years. I am not so sure I would survive in a leadership role in today's BS environment of political correctness and coddling. I hope some of the leadership experiences I faced, both good and bad examples, can assist you in your leadership journey. There is a perception that in the military all you must do in a leadership role is to simply order people to do their tasks. I certainly did see that leadership approach in my Air Force career, but I also observed that the leaders who used true leadership concepts were much more effective.

You very often hear the comment that leaders are born — meaning that leadership is an innate skill; but, in many cases, the leaders of today have been mentored, trained, or modeled into successful leaders. In this book I trace the evolution of my leadership tenets from my time as an ROTC cadet in the mid-sixties through my Air Force career and beyond. I will, where necessary, leave out the specific names of my good and bad mentors to better protect the guilty. This book builds on the leadership tenets that I described in two previous books that I wrote, *Pretzel 06: Memories of a Forward Air Controller in Southeast Asia 1970-1971* and *Top Cover for North America: Defending America from the Soviet Threat*, and highlights the key leadership skills that I developed in times of both war and peace. Some of the leadership situations described in those two books will be repeated in this book to emphasize the real-world application of leadership skills. This book also fills in the other forty-nine years of my leadership development.

The goal of this book is to define the ten key tenets of leadership that I developed in each phase of my personal leadership growth through numerous vignettes about specific examples of those key leadership definitions. Several of these leadership tenets come from some of the very best leaders I have ever observed, and I tried to emulate them in my leadership duties. Other leaders I served under were so bad in their leadership skill set that I wanted to completely avoid the leadership examples they showed. The ten tenets of leadership are:

- Competence
- Mission
- People
- Communication
- Standards
- Integrity
- Loyalty
- Decision Making
- Delegation
- Teamwork

As I present my thoughts on leadership, I will use the analogy of building a house. When I was a young child my father, grandfather, and uncle were house builders in Oregon. One lesson I vividly remember from those experiences was the importance of building a very strong foundation to each house. Once the foundation was purposefully built, all the other components of the house — the walls, the windows, and the roof — would be sound and built on the strong foundation. So in this leadership model I will define the four foundational leadership skills that form the basis for the six additional leadership skills that will follow, thus "building" my leadership model. At the end of each chapter I will list the most important leadership lessons I learned through my hands-on leadership experiences.

1

COMPETENCE

Of all the leadership traits that I have observed over the years, the one that stands out as my number one foundational skill is competence. How many times have I served under a leader who was completely incompetent in leading an organization? Too many times to count. The leaders who gave me the most critical insight into the leadership model were those who were fully competent in their leadership position. That observed competence ranged from combat flying in Southeast Asia to testing air-to-air missiles, from teaching at the Air Force Command and Staff College to leading a fighter squadron, and from being a base commander in Alaska to consulting for a Fortune 500 defense contract research organization; it also includes running a nonprofit program. I was fortunate to work for some very competent leaders and learned valuable leadership skills from each one of them. Competence is one of the four main leadership skills that will be part of the foundation of my leadership model. The stories in this chapter will fill in and explain the competence factor in my leadership model.

ROTC Lessons

The key leadership trait of competence became crystal clear to me when I was in my second year of U.S. Air Force Reserve Officer Training (ROTC) at the University of Portland in the mid-sixties. I was pretty loose and carefree my first three semesters in college. I had put academics on the back burner, but really had my golf game in order. I managed to play golf at least two times a week no matter my class schedule. The problem was that my grade point average (GPA) was a paltry 2.2. The sergeant in charge of the ROTC cadets brought me into his office and explained that without getting my GPA up to at least 2.5 I would not be allowed to continue in the ROTC program for my junior year. That could certainly put me in a position to be drafted into the Army as soon as college was completed because the draft was alive and well in 1966. So, I needed to get my butt in gear and pull my grades up in the second semester of my sophomore year. This dedicated and talented sergeant also told me that he had worked for a few leaders in his career who were less than competent. He did not want me to be one of those "dumb shits," a know-nothing leader, and told me that if I wanted to be an effective officer in the Air Force, I must be the smartest person in any organization I happen to serve in.

The sergeant came up with a plan for me to get my GPA up to the minimum standard and continue in the ROTC program. It would not be easy, but it could be done. He walked me through the classes I needed to take, totaling twenty-one hours of very tough academic course work. He also required me to check in with him two times a week and give him an update on how the classes were going. It was a very challenging load but through a lot of hard work and no golf that semester, I got a 4.0 average on my class work and brought my GPA up to a respectable, for me, 2.8. The study habits and discipline that the sergeant taught me really paid off for the rest of my days in college and taught me the foundational leadership skill to be the very best I could be in any situation I found myself in. This renewed focus on my studies enabled me to earn the Distinguished ROTC Graduate award as well as my commission as a second lieutenant in May 1969. The sergeant also impressed on me to "remember that the most important assignment you will ever have in the Air Force is the one you are currently doing." I carried that philosophy with me for the remainder of my working days both in the Air Force and in civilian life following my retirement from the military.

Undergraduate Pilot Training (UPT)

Attending UPT in Lubbock, Texas at Reese Air Force Base (AFB) was one of my initial Air Force dreams come true. I would actually be training to become an Air Force pilot. While there were one hundred students who started UPT with me in Class 70-07, all but two of us were ROTC graduates. The remaining ninety-eight students were recent graduates of Officer Training School (OTS). Discussion with many of the OTS graduates showed that they were at pilot training because that was the only way they could join the Air Force and not the Army. Very few of them were really motivated to be there, and when my class graduated in May of 1970 there were only forty-five students from the original one hundred who graduated and received their silver Air Force pilot wings.

How did competence show itself in UPT? I was blessed to have one of the most professional aviators assigned as my instructor pilot (IP) for the T-37, a twin engine, side-by-side seating training jet. This IP was always prepared, had answers to every one of my many flight questions, and showed me how to use the visualization concept to prepare for each phase of the flight, including preflight, engine start, taxi, takeoff, maneuvers in the air, landing, and the taxi back to the ramp. He reiterated time and time again to be prepared for anything, to study the procedures, and to remember the lessons learned from all previous flights. He insisted that I visualize the necessary steps in instrument flying (cross check, cross check, cross check) and perform each action necessary to the most precise standards; I can vividly remember him saying, "You are twelve feet too high on your altitude," or, "You are three knots faster than the desired speed," or, "You are one degree off the center line." Learning this precision awareness — this intense level of competence — was a key basis for my future successful Air Force flying career and helped me earn the coveted Skunderdud Award for producing one of the top instrument T-37 flight evaluation scores in the squadron.

This training model ingrained in me by my IP became a valuable guide as I progressed through the remaining months of training to receive my silver Air Force pilot wings. The training visualization technique and absolute goal of perfection on each flight maneuver served me well as I used it in all eight aircraft I would later fly in my

Air Force career. This technique also led me safely through 535 combat sorites in Southeast Asia and enabled me to earn the instructor pilot designation for every aircraft I subsequently flew.

Combat Competence

Leadership can be developed in many different situations. In my particular case, my combat flying time (752 hours) in Southeast Asia was probably one of the most foundational scenarios that built my leadership portfolio. I was just a twenty-three-year-old first lieutenant trying to fly combat and stay alive on my first operational Air Force assignment. The learning curve there was very steep, but I was fortunate to work with an older Australian pilot assigned to our squadron who did my in-country evaluation as a forward air controller (FAC). On every single flight he impressed upon me that to be an effective FAC I had to learn and improve my skills in many different areas. Flying the airplane was really a secondary skill I would need. I also had to know my area of responsibility (AOR), observe even the most subtle changes to the landscape, and understand the capabilities of the Army troops I would be supporting, the capabilities — in great detail — of the elusive enemy we were fighting, and the capabilities of each weapon system that I would be using to support the ground troops. These skills would be learned, fine-tuned, and put into practice on each and every one of my 535 combat sorties that I flew in 1971. This seasoned pilot from Australia insisted that I improve my skills, expand my skill set, and apply lessons learned on every single flight I made.

Part of the FAC's job was to build a set of AOR maps for everyday use. I learned from another FAC the importance of putting together the set of maps that I would be using on every single mission. Remember that this was before the invention of the Global Positioning System (GPS). My task was to assemble the 181 maps I would need for my AOR and have a predetermined plan on how I could easily access and read the maps while flying, being shot at, and trying to control multiple fighters striking the enemy forces that were attacking the friendly ground troops I was supporting. I quickly became the "go to" FAC in our squadron on how to develop the necessary set of maps for each new pilot who arrived in the squadron. Being competent was starting to make sense to me.

In addition, this very talented Aussie pilot taught me the most amazing technique of launching rockets to mark a target that I needed the fighters hit. I learned the technique and practiced it almost daily to the point where I would almost always win any rocket shooting contest against my fellow squadron pilots. With these new set of skills that I worked to refine throughout the year, I was in a position to be the only instructor pilot (IP) in our small group of FACs. Being the most competent FAC in our unit impressed on me the importance of competence as a leadership skill; I tried to develop this skill no matter what my current assignment was.

Aircraft Accident Investigation Competence

My unique leadership training journey took an unexpected change in direction with my next assignment following the combat tour in Southeast Asia. I was assigned as a flight safety officer at the Armament Development and Test Center (ADTC) at Eglin AFB, Florida. My job was to ensure the safe flying operations of all of the armament flight test missions that were flown out of Eglin. In addition to reviewing every flight test plan and preparing a hazard analysis of the test plan, I also had the privilege of flying some of those test missions myself, either as the primary test pilot or as a photo/safety chase pilot. The missions were very challenging and exciting. I also had the additional task of being the flight safety representative on any aircraft accident investigation board. And, believe me, in the flight test business there were multiple opportunities to serve on an accident investigation board.

A very experienced master sergeant in the ADTC Flight Safety Office gave me some wise career advice as soon as I arrived at Eglin: "Make sure you take advantage of every formal training course available for your current position. That additional training will give you the edge in whatever task you are assigned." I took his advice and attended the four-month U.S. Air Force Flight Safety Officer's Course at the University of Southern California. I also attended the one-month Crash Survival Investigation Course at Arizona State University. These two detailed flight safety courses course gave me the necessary initial background experience on the key parts of running a flight safety program and accident investigation board activity.

However, in spite of reviewing every test flight activity at Eglin, one of my main tasks as a flight safety officer at the ADTC was to coordinate and advise on any aircraft accident investigation that was conducted following an aircraft mishap. While I just a captain at the time, the accident investigation boards that I served on as flight safety officer (six of them) were comprised of a colonel as the board president, a lieutenant colonel as the investigation officer, a major or above as the aircraft maintenance officer, a major or above as the unit representative, and a major or above as the medical officer (if it was a fatal accident). Depending on the type of accident, a representative from the aircraft manufacturer provided technical advice.

Since accident investigation was a new field of experience for me, and because I had not yet served on an aircraft accident investigation board, I took time in my daily tasks to apply what I had learned in the flight safety courses by reviewing the previous six accident board investigation reports. I scoured them from front to back to glean whatever information I could on the process, the format, and the methodology used to arrive at a conclusion as to the cause of the accident and the recommendations to prevent a similar accident from happening. I did find numerous typos in the reports and made a pledge to myself to make zero typos for all aircraft accident investigation boards that I may be a part of in the future. I also noticed in a few of the accident reports that, while the findings of the accident board made sense and seemed to be proven in the text of the report, there were a few very large leaps of logic getting to the findings. I also made a pledge that I would ensure that the flow and logic of the findings made sense to the reader and that the actual accident board findings followed the logic in the content and format of the report. In other words, I wanted to make sure even a fighter pilot could understand the board's logic and findings. I am happy to say that not a single error was found on any of the six major accident investigation board reports that I served on as the safety advisor.

I also had the task of leading and presenting an annual Aircraft Accident Investigation Board Training Course to the pre-selected unit members who would be called upon to be a member of an accident board if needed. I was initially reluctant to do this training because I had yet to serve on an accident board. Once again, the wise master sergeant in the Flight Safety Office gave me some very timely advice. He recommended that I build the course outline, schedule the training, and coordinate multiple guest speakers who had actually served on accident boards to give the specific details of their particular area of expertise. That approach was a complete success, and I received positive feedback from the class attendees. However, it was not long before I was called to be the board advisor on two consecutive accident investigations, and I rapidly became the expert at Eglin on the conduct of the accident investigation board process.

I have two interesting stories about incidents that occurred during these two investigations. The first incident was during the investigation of a fatal F-4 accident where the backseat pilot was killed upon ejection. He had ejected over the Gulf of Mexico, and we did not find the body for several weeks. Finally, after many search operations a partial body washed ashore several miles from the base. The accident board president asked me to drive to the location about seventy miles away where the remains were found and bring them back to the Eglin base hospital for processing. I drove the ADTC Flight Safety Office vehicle to the location and found that a partial flight helmet with human remains had been found washed up on the beach along the Gulf of Mexico. I rummaged through the safety vehicle and found a two-pound coffee can that held some of our safety supplies. After carefully placing the human remains in the coffee can, I drove back to the base hospital. Having never processed human remains before, I went directly to the Eglin AFB hospital emergency room only to find that it was overflowing with patients and family members waiting to be seen.

As I stood there waiting, the medical tech behind the desk said, "Captain, it will be a few minutes before we can see you." That was not the right answer for me, and I said in a very loud voice, "I have human remains in this coffee can, and I need someone here immediately who can take care of this issue." Wow! I have never seen a medical staff member move so quickly. I was immediately escorted to a treatment room and had the attention of an Air Force doctor — a colonel — in less than three minutes. This particular accident was caused by the failure of a fairly simple hydraulic actuator, making the aircraft uncontrollable.

The second incident involved another F-4 accident where both pilots were able to safely eject. The front-seat pilot, however, was severely injured. The accident board convened, and the investigation team got the process started. Since both pilots had survived the accident and subsequent ejections, it was important that we collect statements from them to help us guide the investigation in the right direction. The board president asked me set up an interview with the front-seat pilot so we could start the investigation. Fortunately, my wife was an Air Force captain at the time and was the charge nurse on the floor where the injured pilot was being cared for. I told the board president that there would be no problem with the interview, and we both headed to the hospital. Once there, my wife informed me that her patient, the pilot involved in the accident, was in no condition to be interviewed at the time. We would have to come back in a few days when the pilot was in a better medical condition. The accident board president asked me if I was sure that it was my wife who would not let us interview the injured pilot. Several years later, this same pilot became one of my upper-level supervisors and always thanked my wife for saving him from the accident board interview interfering with his recovery. This accident was caused by a lightning strike on the aircraft which caused the fuel tank in the left wing to explode and, obviously, made the aircraft unflyable.

Flying Competence

As I said earlier in this book, I had the privilege of flying many of the flight test missions while assigned to the Armament Development and Test Center at Eglin. I had become an instructor pilot (IP) in the C-131, the T-38, the T-33, and the T-39 aircraft, and I flew many test missions and training flights. The C-131 was a two-engine (reciprocating engine) military cargo plane that could carry up to forty-eight passengers, had a crew of three, and could fly at 240 miles per hour up to 24,000 feet. The T-39 was a two-engine, mid-sized business jet aircraft that was piloted by a crew of two, could carry up to seven passengers, and could fly up to 550 miles per hour at 40,000 feet. One day my operations officer told me he wanted me to check out in the T-39 and become an IP as soon as possible in that aircraft as the Armament Development and Test Center commander, a two-star Air Force general, requested an experienced instructor pilot accompany him whenever he flew. His request was for a young pilot with combat flying experience. This general had combat flying experience in World War II, Korea, and Vietnam. So, I began the very quick task of getting checked out as a pilot in the T-39 followed the very next day with my checkout as an instructor pilot. In order to get the instructor pilot flight check, I needed at least one hundred hours of flight time in the T-39. The operations officer told me to find another T-39 IP and get the one hundred hours of flight time as soon as possible. With some good weather and a very reliable airplane I was able to complete the required hours in just under three weeks. The next day I completed the T-39 IP check ride with a "Highly Qualified" rating and two days later had my first flight with the general. By the way, the general was an excellent pilot, and I no issues flying with him on thirty-plus trips around the country. Whenever we encountered a hairy flight situation he would say over the intercom, "You have the aircraft," and I would take over the flight.

Self-Help Competence

While I was doing my duties as an operations officer with the PQM-102 drone program at Holloman AFB in New Mexico between 1977 and 1980, I decided that it was time to enhance my chances for promotion to major as I had been in the Air Force for eight years and knew the majors' promotion board would be convening for my year group in the next eighteen months. I signed up to complete Air Command and Staff College (ACSC), attending the seminar program two hours each week for the next fifteen months. It was a fairly easy course, but it did require attendance at each seminar session, multiple tests, a comprehensive final exam, and completion of a large research paper to pass the course.

Simultaneously with the ACSC course, I also enrolled in a master's degree program in business administration (MBA) with Pepperdine University's on-base program. This program was a difficult and time-consuming endeavor. Our class met every week on Thursday and Friday evenings from 6:00 p.m. until 10:00 p.m., on Saturday from 8:00 a.m. until 5:00 p.m., and on Sunday 8:00 a.m. until 2:00 p.m. for one entire year. The course consisted of twelve different classes with each requiring a midterm exam and a twenty-five-page paper, and then an eight-hour comprehensive final exam on all twelve subjects. Needless to say, I was a very busy person during this time. I was still continuing to do my operations officer job and was fortunate to fly at least three times a week. I completed the ACSC program and finished my MBA before the promotion board met. This hard work at enhancing my competence paid off as I not only earned a promotion to major but was also selected to attend ACSC at Maxwell AFB later in my career; in-person attendance at ACSC is only awarded to the top 30 percent of newly promoted Majors.

Outside Your Comfort Zone Competence

One of the more unique turns in my Air Force career involved a significant move to a mission that I had very little hands-on experience. I was assigned as a senior flight commander in an F-106 fighter interceptor squadron (FIS) at McChord AFB, Washington. The F-106 was a single-seat, all-weather fighter that had a crew of one and a max speed of 1,525 miles per hour (Mach 2.3). It could carry four AIM-4 radar-guided missiles, one AIR-2A Genie nuclear rocket, or a single M61 Vulcan rotary cannon. I truly enjoyed that job because I was able to fly many times each week. Our squadron, the 318th FIS, had just undergone a very intense operational readiness inspection (ORI) from the headquarters of the Tactical Air Command. While the operations side (flying) of the squadron did quite well on the ORI, our aircraft and weapons maintenance sections failed the inspection miserably. Our squadron commander, a lieutenant colonel, was fired on the spot, and a new commander arrived to "clean up" the squadron.

During the first week of the new command, the boss called me to his office. I assumed that he was going to make me an assistant operations officer in the squadron. What a surprise when he decided to put me to be in charge of the component repair branch (CRB) in the squadron. The CRB consisted of more than 255 enlisted maintenance members and three officers tasked with the back-shop electronic repairs, the life support maintenance section, the parachute shop, and the storage and maintenance of the nuclear AIR-2A missiles. I tried to protest, but he said, "What the CRB needs is leadership, and I believe you are the one to lead the CRB back to a satisfactory status." I did get a concession from the commander that I could keep actively flying the F-106 as long as I made regular progress leading the maintenance team in the right direction.

My first task was to find the smartest senior enlisted person in CRB and start picking his brain about the maintenance unit. What amazing insight he provided. The basic problem in CRB was the maintenance personnel had no direction nor insight about how their individual tasks contributed to the overall success of the squadron. I immediately set up a schedule to visit every work section in CRB to gather information and input about their tasks, their limitations, and, just as important, their needs — anything that would impact their ability to keep the F-106s flying. I quickly got up to speed on the issues I needed to address as their new leader.

In order to clearly demonstrate the important relationship between back-shop maintenance to the ultimate task of keeping the F-106s flying, I made a point to call the specific mechanic who had worked on any part that failed during any of my flights, explaining why I had maintenance issues with the aircraft. I wanted each one of my maintenance technicians to see the exact relationship between their hands-on, back-shop maintenance work and the pilot's ability to fly the airplane for a successful sortie.

In addition, I made it a point to visit, for at least one hour every week, each of the seven support shops in CRB. I wanted first-hand data on those areas that needed the most emphasis. I quickly learned that some of the shops were just going through the motions without realizing how their individual activities related to the squadron's overall mission. I also made a point of getting the daily CRB maintenance summaries at least two hours prior to the morning squadron staff meeting so that I could study the reports and make myself smart on the data points on each area we were tracking.

This daily review of the maintenance status reports demonstrated my leadership of the CRB and gave the squadron commander and the maintenance chief the confidence that I was leading the CRB in the right direction. Following a very intense six-month learning process for me, I implemented a successful change in direction, and the CRB passed the follow-on operational readiness inspection with flying colors, receiving an "Excellent" rating.

My maintenance competence process did not stop with the ORI. Since I was now considered a maintenance officer as well as a pilot, I took the advice of the former safety office master sergeant to "get all the formal education you can." I insisted that I attend the Senior Officer Maintenance Officer course at Chanute AFB in Illinois. The four-week long course covered much of the information that I learned hands-on in CRB, but it also gave me the formal Air Force designation as a qualified aircraft maintenance officer. This designation became important later in my career when I was a flying squadron commander and a base commander. The formal aircraft maintenance training gave me great insight and credibility into how the fighter wing maintenance team supported our squadron's flying and alert mission.

Outside My Comfort Zone Again

After promotion to colonel, I was named the Deputy Director of the North American Aerospace Defense Command (NORAD) planning shop and lead a twelve-man staff that wrote and implemented all NORAD plans and logistic support. I was very comfortable in this position as I had a lot of experience at all levels of the NORAD mission in the field. One of the newest and most technologically important programs for NORAD was the Over-the-Horizon Backscatter Radar (OTH-B) system that was just being implemented to help track Soviet air activity trying to enter our airspace as well as drug trafficking aircraft flying from Central and South America delivering vast quantities of illegal drugs into U.S. The OTH-B radar had the capability to bounce its radar signal off the earth's ionosphere and then receive that "bounced back" data to track a single aircraft thousands of miles away. Since I had been the division chief for NORAD's counter-narcotics planning mission before I was promoted to colonel, I knew a lot about the trafficking attempts to fly illicit drugs into our country.

One of the main obstacles to the new OTH-B radar program was a congressman from New York who was trying to defund the OTH-B program because he wanted some additional Air Force National Guard C-130 aircraft to be assigned to his New York congressional district. This congressman wanted to visit the OTH-B radar sight in Bangor, Maine so he could have a better understanding of how to kill the program. Our four-star general decided that NORAD needed to be represented at that visit to convince the congressman of the importance of the OTH-B radar program to NORAD's surveillance capabilities, especially in the counter-narcotics world. I was selected to be the NORAD representative at the congressional visit.

While I had a fighter pilot's knowledge of the OTH-B program and could describe the big picture of the operational use of the OTH-B in NORAD's overall surveillance mission, I knew I could very easily get over my head with some of the technical questions that might come up. As expected, one of the congressman's assistants was well prepared for the visit and asked some very technical questions about the OTH-B Radar system. Fortunately, I had coordinated with the colonel who commanded the OTH-B site to answer any technical questions even though I was the senior NORAD representative at the meeting. This colonel answered each technical question perfectly and the OTH-B radar program was saved for another two years. The competency lesson I learned here was to always trust the experts when you don't have all the answers.

Competency Lessons

Following my successful Air Force career I retired from active duty in November 1995. The need to be competent did not stop when I retired. My initial goal upon retirement from the Air Force was to do anything but defense contract work. I just didn't want to get into that game and use my influence to sell stuff to the government that they didn't need. So, I resisted as long as I could. An offer did come along that I could not refuse as I had the first of my three children just starting college, and I needed the funds to help them out with tuition. I was contacted by a company near Ogden, Utah to be a consultant for a couple of very large firms that did business in Colorado Springs. The support I would provide would be for the Air Force Intercontinental Ballistic Missiles (ICBM), satellite development, and satellite control system programs. These new areas of consulting were not in my competency comfort zones as I never had any operational experience in these three areas. The offer, however, was just too lucrative to pass up. I also did many hours of additional consulting work for other various companies from flying support at the U.S. Air Force Academy to working on numerous contract proposal reviews to reviewing the future plans of various military organizations.

But off I went to meet the main client in Boston for a job interview. I was certainly concerned about my very weak résumé in those three technical areas: ICBM missiles, GPS systems, and GPS support. As soon as I sat down with one of the division chiefs in charge of satellite control systems, he literally threw my résumé across the desk to me and said, "I don't see any hands-on experience in satellite control systems. How can you help me?" My answer was very direct. "You are correct," I said, "I don't have any experience in this area, but I do have some unique experience not everyone has and that is direct access to the Air Force leadership that makes all the decisions concerning satellite control systems."

My last position in the Air Force was as the Deputy Chief of Staff of the North American Aerospace Defense (NORAD), and I had worked directly for a three-star Canadian general and a four-star U.S. Air Force general. I had a working relationship with both of those leaders — plus the entire senior leadership staff of NORAD and Air Force Space Command. The interviewing division chief's answer was straightforward. "OK we will give you chance and see what you can do for us."

I had to get my butt in gear and learn as much as I could about these three systems. Fortunately, I knew a lieutenant colonel in Air Force Space Command who had worked for me as a major and was a satellite and ICBM expert. He agreed to become my mentor and translate these technical areas into something a fighter pilot could understand. His diligent work with me, plus endless hours of self-study on my part, made me just enough of an "expert" to become dangerous. The client was satisfied with my performance, and I was a consultant with them for more than thirteen year. Many times they would send me solo across the country to attend some of the most technical conferences with the expectation that I would capture the needed material to help them make a wise decision on future programs to support. I learned that competence at every level in the leadership model is a critical skill.

Volunteer Competence

I also earned a valuable lesson about the need for competency from a volunteer program I have been involved with for almost twenty-five years. The first month following my Air Force retirement my pastor asked me to attend a week-long training session in San Diego to learn how to set up an all-volunteer ministry program. The program, called Stephen Ministry, is a Christian lay ministry program that helps people going through tough times in their lives. The volunteers complete a very intense fifty-hour training program before they are commissioned to provide help to those in need; the volunteers are also supervised by volunteer leaders who have served for several years and who have completed additional leadership training.

My primary role in this ministry was to coordinate all aspects of the program from recruiting and training to supervision, referrals, and program support. I taught a good portion of the training program, educating nearly seventy-five ministers in the first six years. I always had a strange feeling in my gut as I was training new ministers on how the program actually worked because I myself had not been assigned to a care receiver for the first six years of the program. At the time we thought that it was best not to assign one of the ministry leaders to a care receiver as it would be too much of a burden on the volunteer leader.

That all changed dramatically when we had a very sensitive and difficult ministry assignment that none of our volunteer ministers felt they could handle. So, with some reluctance I agreed to take on this difficult case. The lessons that I had tried to teach the other ministers during their training session became very real to me and really enhanced my ability to "fill in the blanks" on how this ministry worked and gave me the necessary competence and skills to honestly teach the new ministers how the program really works. By the twenty-fifth year of the program I lead, our team had trained more than 175 volunteer ministers.

I was blessed to walk with my one of my care receivers during a very difficult time for almost nine years and learned a lot of the actual hands-on approach to this important ministry. I was now very competent in this skill of lay ministry and learned that no matter what you do in life being competent is a critical piece of any leadership formula.

Competency Leadership Lessons
- Always do your best, no matter what you are doing.
- Learn all you can about your mission/job.
- Complete as much formal education as available.
- Practice, practice, practice the best techniques.
- Find the best people.
- Learn every detail of your assigned mission/job.
- Find and follow a role model.
- "Stretch" your competence portfolio.
- Trust your experts when you don't have all the answers.

2

MISSION

Following leader competency, the second most important factor in my leadership philosophy is mission. Members of the Air Force and other military branches are charged with specific missions for each assignment, whether they wear a uniform or are a civilian employee. From my Air Force career, my missions included flying combat, testing air-to-air missiles, investigating aircraft accidents, monitoring government flying contracts, sitting air defense alert, leading an Air Force maintenance unit, teaching at the Air Command and Staff College, directing the readiness of an entire command, commanding a tactical flying squadron and serving as a base commander in Alaska, leading a counter-narcotics planning division, and serving as the deputy chief of staff of a very large multinational organization. In my civilian life, my missions included working in the defense industry and leading a volunteer ministry program. For each mission I had an important leadership role to play. Some were certainly more exciting and sexier than others, but each mission I was assigned contributed to my development as an individual and as a leader. While assigned to a particular mission, that was the focus of all my activities each and every day. As long as I stayed focused on my assigned mission, I was earning my pay. Some of the missions I was assigned to were not my first choice but, nevertheless, they were my assigned missions, and I needed to do my best in every situation.

Former Air Force Chief of Staff General Curtis E. LeMay stated, "No matter how well you apply the art of leadership, no matter how strong your unit, or how high the morale of your men and women, if your leadership is not directed completely toward the mission, your leadership has failed."

As a leader in each one of these missions I felt that it was my role to instill the concept of "mission first" to everyone I came in contact with, whether I was a formal or informal leader in that organization. This was especially true during my two squadron commander assignments. The commander is always expected to guide the entire squadron and focus the energy of the squadron members in the right direction. Everything else was not important.

In my civilian roles I felt the same duty to mission to give the best advice to my clients and to give the best training and support to the volunteers in my ministry program. I will now outline some of those key missions that I had in my life and describe the leadership qualities I found in so many diverse areas.

Combat Missions

One of the first real-world missions that taught me so much about leadership was my assignment to the war in Southeast Asia as a forward air controller (FAC) in 1970-1971. The FAC's mission was to fly directly above the battlefield in a small, unarmed observation aircraft to find and mark the enemy targets with a white phosphorus spotting rocket, to control the fighters that dropped the bombs, and to do a bomb damage assessment after each attack. The FAC's expertise as an air strike controller also made him an intelligence source, munitions expert, communications expert and, above all, the on-scene commander of the strike forces and the start of any subsequent search and rescue mission if necessary. Since the FAC was assigned to a specific geographical area, he became the absolute expert on what was happening in that area of responsibility (AOR). The FAC knew whenever a blade of grass was misplaced or that some activity had taken place in the AOR. Whenever a FAC supported U.S. troops on the ground the FAC had to be designated an "A" FAC and have some fighter pilot qualifications. That was the designation I earned by completing the Fighter Lead In program at Cannon AFB in the fall of 1970. This requirement was in place to ensure there were no friendly fire incidents killing U.S. troops on the ground with an errant bomb or missile. Protecting the friendlies on the ground from the ordnance that the FAC controlled was one of the most precise and heart-stopping aspects of being a FAC. The strike missions we were expected to support were either preplanned strikes (scheduled at least twenty-four hours in advance on a specific target) or an immediate airstrike when an urgent situation on the ground required air support. In my case, I had the additional responsibility of inserting, supporting, and extracting Special Forces teams and a very intense High Low visual reconnaissance mission. We could be expected to do some or all of those tasks on each mission. We flew at around 1,000 feet or lower above the battle so that we could find the targets, observe the action, and act as easy target practice for the bad guys. There was a North Vietnam and Vietcong philosophy about shooting at the FAC: the bad guys never wanted to start shooting at the FAC until the fighters were on scene. I guess they had the hope that if they didn't shoot at us it would be harder to find them. But once they knew we had detected them it was like an arcade shooting gallery. Toward the end of my time in Southeast Asia the bad guys added a very fearsome weapon, the hand-held SA-7 Strela surface-to-air heat-seeking missile. This weapon was in addition to the AK-47 assault rifle and 12.7 mm (.51 caliber), 14.7 mm, and 37 mm anti-aircraft guns they used throughout the war. I had many close calls, and many rounds hit the aircraft I was flying. We used to call those hairy missions "seat sucking missions" which meant that after we landed we needed help from the crew chief to pull the seat cushions out of our asses. During night missions we flew with our exterior lights off to lessen the chances of being shot down. The most imminent danger during a night mission was a mid-air collision with the fighter aircraft we were controlling because they also flew with their lights off. Night missions were doubly exciting as we could see the many tracers that were fired at our aircraft. The NVA/VC usually had about every fifth round as a tracer round. The tracer rounds helped them see where their fire was going but also showed us exactly where the fire was coming from. I occasionally saw tracer rounds during day missions.

We did have a few other sophisticated techniques to keep from getting shot down: trying to make our attack with the sun at our backs if the terrain and sun angle were right, flying behind a ridge line if the terrain was good, and, best of all, flying with one foot on one of the rudders. This full-rudder deflection made us fly almost sideways to give the impression from the ground that the nose of the aircraft was tracking one way while in fact we were flying about thirty degrees off that track. It was amazing to see the tracers track way out in front of our aircraft. Sometimes the more experienced gunners found out about our trick flying, and many of my fellow pilots were shot down. We also used a jinking maneuver — a very uncoordinated and erratic flight movements — that never gave the enemy a predictable flight path. I tried to never fly a straight flight path more than a few seconds.

We had one other cool technique we used to protect us. The squadron sent down a directive that we must fly with our flack vests. Piloting an aircraft while wearing a flack vest, a survival vest, and a parachute was just too cumbersome so we would place our flack vests under the seat cushions. Our thought was at least we would save the family jewels if we were hit in that part of the aircraft... all of us thought this was a brilliant idea. For me, I was still alive and felt invincible.

This combat time gave me some very detailed ideas about leadership. Certainly, being in control, making split-second life or death decisions, owning up to decisions, and getting better each day at carrying out the mission taught me the basic skill of leadership that I would subsequently apply to whatever I was assigned in my military or civilian life.

Flight Safety Mission

Following my one-year combat flying tour in Southeast Asia I was assigned to a unique mission at Eglin AFB, Florida. Based on my experience controlling aircraft during the war, my Air Force Specialty Code (AFSC) was a match for an assignment to the Armament Development and Test Center (ADTC). This was certainly not my mission of choice as I would have much rather been assigned to a fighter squadron flying every day. But this was the mission I was assigned to, and I tried to do the very best in every aspect of the mission. The mission of the ADTC was to test all kinds of air-to-air and air-to-ground munitions. My role to support that mission was in the Flight Safety Office, and I reviewed and approved munition test flights in addition to conducting aircraft accident and incidents investigations. I described some of the competence needed for this mission in the first chapter of this book.

What a change in missions for me. During the war I was at the pointy end of the combat stick controlling air strikes against the enemy. At ADTC I took on a much more analytical role and really had to use some brain power to contemplate all the potential safety issues that could arise during any particular test flight. As a young captain I questioned if I was ready for this level of responsibility. Fortunately, I had three very experienced lieutenant colonels in the Flight Safety Office who really mentored me and got me headed in the right direction with this important new mission. The learning curve was very steep but with some direct guidance from my mentors I was able to give highly detailed safety reviews of the proposed flight tests that I approved.

One aspect that helped me learn my new mission were the many Operation Safety Analysis (OSA) reports on file that covered just about any type of flight test activity conducted on the Eglin test ranges. I made it a priority to review almost every OSA on file and gain some insight into how I would then evaluate and approve any flight test plan that I reviewed. During my three years in Flight Safety not a one of my OSAs was ever questioned or suspected in any flight incident.

Another factor in my mission learning process was the opportunity to fly some of the flight test missions that I had approved, either as the test pilot or flying as the photo/safety chase pilot. I was able to get firsthand experience in the test business and found that flying these various test missions improved my competence in safety analysis.

One unusual accident investigation I was involved with led to an interesting follow-on assignment from Eglin. When I first arrived at Eglin I was assigned to the DT-33 aircraft that supported the QF-104 unmanned drone program. The QF-104 was a single-seat, single-engine, unmanned, full-size F-104 fighter aircraft converted into an aerial target for high-level air-to-air missile tests. It could fly Mach 2.2 at 50,000 feet. The DT-33 was a highly modified single-engine jet training aircraft that had some unique capabilities. Once the unmanned QF-104 was launched the DT-33 would join in close formation with the drone and electronically take over control of the drone aircraft. The front-seat pilot of the DT-33 had twenty toggle switches mounted on the top of the front cockpit to change and manage the engine speed and all the control surfaces of the QF-104 while the pilot in the backseat actually flew the DT-33. Once the drone was over the Gulf of Mexico and in the designated test airspace a ground controller would take control of the drone and fly it through the pre-established flight profile. If the drone survived the missile firing at the over-water test range, the QF-104 would be remotely flown back to the shoreline where the DT-33 would once again fly in formation to gain control of the drone and fly it to a final approach position about ten miles from the runway. On final approach personnel in a mobile control station would take control of the drone and fly it to a safe landing. This entire drone operation was a complicated set of coordinated flying by both the ground controllers and the drone controller in the DT-33. While I was only on the periphery of the QF-104 drone program as a support pilot I learned a great deal about drone operations and the risks involved in this important test mission.

In 1973 a new drone aircraft, the PQM-102, was being brought into the program to replace the QF-104 drones. The F-102 was also a full-size, single-seat, single-engine fighter that could fly at Mach 1.0 at 50,000 feet. The aircraft flew out of the Crestview, Florida airport, just north of Eglin AFB. Crestview was one of the sites where the F-102 was being converted into the PQM-102 drone version. One day the landing gear on one of the PQM-102s collapsed on landing at Crestview. While the aircraft was not destroyed there was significant damage to the aircraft, and an accident/incident investigation was ordered. I was appointed as the incident investigation officer for this incident and traveled to Crestview for two weeks to review the operation of the PQM-102 as well as the varied support activates of the program. While I determined that the incident was caused by a broken landing gear sway bar, I did find many unsatisfactory maintenance and supply issues with the Crestview PQM-102 operations. I completed my report and passed my recommendations along to the Test Wing personnel who were conducting the PQM-102 program. Almost by default I became the PQM-102 drone expert in the Test Wing.

Exactly six months after submitting my PQM-102 incident report I was advised that I was going to be transferred to Holloman AFB in New Mexico to become the assistant operations officer and government flight representative of the PQM-102 drone detachment stationed at Holloman. The Holloman PQM-102 Detachment was a sub-unit of the Air Defense Weapons Center (ADWC) at Tyndall AFB, Florida and would be supporting the drone operations on the White Sands Missile Range that was located adjacent to Holloman AFB. In addition to the ADWC assignment, I was also attached to the 6585th Test Group at Holloman as the head T-38 instructor pilot and flight examiner for the Test Group. I had hit another jackpot as I would be flying many times each week either supporting the PQM-102 program or flying test missions in support of the 6585th Test Group.

As a government flight representative (GFR) I did more than fly in support of the PQM-102 Drone Program; I was also the government representative for all drone operations including the range safety for every launch and recovery of the drones. GFR duties also included monitoring the PQM-102 contract to ensure all work was being completed by the exact letter of the contract. The leadership lesson I learned here was to become the expert on exactly what the contract called for and exactly how the contractor accomplished the tasks as outlined in the contract. I learned the importance of the "black and white" interpretation of a contract and the importance of ensuring the government was getting its money's worth from the contractor. This valuable lesson became very important in a future Air Force assignment I had following my time at Holloman.

Fighter Pilot Assignment

My next mission involved serving as squadron pilot in an F-106 fighter interceptor squadron (FIS) as the senior flight commander at McChord AFB in Tacoma, WA. My primary mission was to fly, support air defense alerts in Tacoma and at Kinsley Field in Oregon, and make sure the six pilots in my flight all completed their required annual training requirements. It was a great thing to be just responsible for the flying mission and the actions of six other professional pilots. I really enjoyed the pace of activities but that didn't last for long. All the "just flying mission stuff" changed dramatically for me when our squadron failed an operational readiness inspection (ORI) due to many wide-spread maintenance issues. I explained this process in Chapter One. I was less than pleased to be a new maintenance officer but that was the mission I was assigned. I needed to do my best and provide the leadership skills necessary to raise the maintenance function back to a satisfactory level.

Conversion Program Director

If being in maintenance was not enough to stretch my mission expertise, I was also assigned as the squadron program director to convert the 318th Fighter Squadron from the F-106 Delta Dart to the new F-15 Eagle aircraft. This new mission really impacted my ability to maintain flight currency to fly the F-106 and lead the 255 maintenance folks in the CRB at the same time. While it was a challenge, I managed to maintain my aircraft currency right up until the time I was reassigned to the Air Command and Staff College (ACSC) in Montgomery, Alabama.

The conversion process involved the entire spectrum of activities to include new facilities, pilot training for the new aircraft, supply support, and the maintenance support assignment process. Because we had a higher headquarters co-located with us on McChord AFB, this mission required some very delicate high-level "political" maneuverings with senior staff who thought they knew more than I did about the conversion. Every time I briefed the higher headquarters staff, I had to have rock-solid answers to the many questions posed by the staff. The conversion program was a complete success, and the squadron became proficient in F-15 operations in the minimum time allotted.

Faculty Instructor

Following my very intense eighteen-month assignment at McChord Air Force Base I was assigned as a student to the U.S. Air Force Air Command and Staff College (ACSC). The goal of the ACSC was to prepare field grade officers — typically majors in rank — from all military branches to serve as staff officers or commanders. In the mid-eighties the normal attendance at this ten-month long college was nearly 700 students. We also had more than eighty students from various countries around the world. The curriculum was very low key in relationship to the prior assignments I had completed, and it was a nice break from the twelve-hour days I had been putting in. I made it a point to put my best effort forward in this situation and decided that I would give a good eight hours each and every day preparing for class, studying for exams, and preparing my class project. That was my current mission, and I gave it my all.

The hard work paid off as I was designated a distinguished graduate and was assigned to the ACSC as a faculty instructor (FI). As an FI I lead classroom discussions, taught an elective course called Roles and Missions of the Air Force, and acted as faculty advisor on many of the detailed research projects each student was required to accomplish. I was fortunate to have four international officers from Canada, Saudi Arabia, Israel, and Peru in the seminars that I led.

Luckily for my career, the commandant of ACSC gave me a concession that he would only keep me on the faculty for two years — the normal assignment for a FI was three years — and then get me back to the operational Air Force. For this concession I agreed to play on his ACSC softball team. I was a fair softball player in those days and figured I could do the instructor mission for two years without jeopardizing my operational experience. During this two-year faculty assignment, I was able to travel to the Canadian Forces Staff College in Ottawa to teach Air Force leadership concepts to the Canadian and other international students attending their college. I also traveled to Bracknell, England and Hamburg, Germany to teach at both the United Kingdom and West German staff college equivalents for one week each. What a career broadening experience those two trips turned out to be as I had many interactions with military officers from those three countries later in my career. The ACSC commanding general was good on his word, and after two years I was able to get back to the operational Air Force with a staff assignment to Alaska. I was also good on my word, and our ACSC softball team won the base championship the very day my household goods were packed up for my assignment in Alaska. My dear wife stayed at home with the movers while I played in a double header.

ACSC Spouses Program

This title may be a head scratcher. One of my most unique missions while a faculty member of ACSC was to be the overall coordinator of the ACSC Spouses Program, one of the more visible support programs at the staff college. Since the normal ACSC class session lasted for more than ten months, most of the students brought their wives and children with them to Maxwell AFB. The Spouses Program presented the spouses with a wide range of military and other information to help them understand and prepare for their military members' future command and staff assignments. There were four distinct groupings of students, called a wing, at the staff college, and a spouse program coordinator supported each wing. I was the overall coordinator for the entire program and had my hands full trying to schedule and deliver the various spouse presentations we gave multiple times each month. This mission, technically an additional duty, was obviously outside my normal Air Force operational experience but I tried to approach this much like I did with all my previous assigned missions: I treated it as the most important assignment/mission I would ever have.

Mission Leadership Lessons

- Your current mission is the most important mission you will ever have.
- Do your very best in each assigned mission.
- Integrate your individual mission into the overall mission of your organization.
- Every part of the mission, however minor, is critical to the overall success of the larger mission.
- Be the subject matter expert in all phases of your mission.
- Pay attention to details.
- Practice, practice, practice.
- Expand your mission comfort zone.

3

PEOPLE

While this seems like an obvious piece of the leadership model there are many leadership facets that need to be explained. People are obviously important in the leadership process and form the third foundational tenet of my leadership model. Each organization that you lead will have people, in some cases many people, who will affect your leadership approach. Remember that my first leadership concept is competence. The people in the organization, to a great deal, will determine the level of competence a leader requires to be effective. I would be less than honest if I said that I had all the pieces of competence to lead an organization. Finding and nurturing the right people can enhance the quality of leadership the formal leader brings to any organization. Realistically a top-notch leader will not have all the expertise needed to run an organization. Leadership examples can come from all levels of an organization, from the very top to the very bottom and everywhere in between. Leadership examples are everywhere. A good leader should seek out those examples and, where appropriate, use those examples to enhance his or her own leadership skills. A corollary to the "great people" part of leadership is that you may find a leader in your organization who is so bad that you want to be the exact opposite in your actions based on the chaos a bad leader can cause at any level in an organization. The following scenarios will highlight the key role people can play, both good and bad, in any organization.

Fighter Squadron

During my eighteen-month assignment in the 318th Fighter Squadron at McChord AFB, I had many areas of expertise I needed to be proficient in. Flying was my best competence but there were more areas I need to be competent in. The very day I was transferred to the squadron maintenance section I found one of the most talented senior sergeants I had ever served with. He was straightforward, knew all the skeletons in the unit, and was not afraid to give me advice, especially in the areas I was weakest. One day he relayed to me that he thought there was a very real problem with the leadership in the parachute shop, which was managed by a senior technical sergeant. His concern was that the sergeant was driving a very expensive car and was wearing too much gold jewelry to be normal; my senior sergeant thought there was something nefarious taking place in the parachute shop. I am glad that I listened to him because one month later the Air Force Office of Special Investigation (OSI) charged this parachute shop technical sergeant with drug trafficking.

In addition to the inside information this talented senior sergeant provided me, he was also an expert in many other areas. One particular skill was his attention to detail. I selected him to be on the F-106-to-F-15 conversion team that I mentioned earlier in the book. One of the main tasks of the conversion team was to lead a quarterly Site Activation Task Force (SATF) meeting summarizing all the actions taken to date and the actions still needed to complete the conversion process. This sergeant was so particular in preparing the SATF reports that he secured the help of three administrative experts in the squadron to help. The reports were usually 300 to 400 pages covering many technical issues. He insisted that there would not be a single typo or misinformation in any of the SATF reports. I appreciated that approach as I used the same requirement in my accident reports in a previous assignment. He came through on his desire for a perfect report. Not a single typo was found in any of the four quarterly reports we prepared.

Another key area of expertise he provided was his awareness about the kind of questions that could arise as I briefed each step in the conversion process to officers from our higher command. I was able to really dazzle the headquarters staff from Virginia during a conversion update progress briefing. This dedicated sergeant had found out in advance that the one-star general attending the briefing had a plan to put the F-15 simulator into an already existing building to potentially save millions of dollars in the conversion budget. The problem with the general's plan was that the building he proposed did not have a wide enough opening to accommodate the simulator. A massive amount of money would be required to open the "nuclear protected" building to accommodate the simulator. In addition, the building proposed by the general did not have adequate power to support the simulator.

My brilliant sergeant had made several overhead slides detailing the problems with the general's proposed plan. As I whipped out the backup slides, I could see the general's face getting redder and redder with each slide. I continued with the presentation and heard no comments from the general. Our original plan to construct a stand-alone building for the F-15 simulator was eventually approved. I learned a valuable leadership lesson with this experience: depend on the experts, no matter where they are in the chain of command, to give you the best advice.

The conversion team earned recognition for their outstanding work with the entire planning process. This was due in large part to my brilliance in selecting this talented sergeant to be on our team. The conversion from the F-106 to the F-15 was completed on schedule, and the squadron had no unnecessary down time in the air defense mission.

Aerial Photo Support

One of the primary flying tasks that I had while in the 6585th Test Wing at Eglin AFB and Holloman AFB was flying photo chase for many of the munition test missions. It was a real challenge to fly at 350 knots while pointed seventy degrees to the ground and keep a steady flight angle so the photographer in back seat of the chase aircraft could get quality photos of the test drop or test firing of the munitions. We had a senior master sergeant who was absolutely the best aerial photographer in the Air Force. He had more than 3,000 flight hours of photo support time. I was always so pleased to know that he would be the photographer in my back seat on any test flight I was assigned to chase. We always had the optimal photo results when he was the photographer. He would always tell me exactly where I needed to position the airplane so he could take quality photos or video of the test mission. I was so confident in his abilities that on a few occasions I would let him preflight the photo chase aircraft if I was running late for the mission.

We had a near tragic incident on one flight in Florida during a very difficult air-to-ground munition test. The mission called for the release to be at seventy degrees to the ground, and the photo chase aircraft needed to maneuver into an exact potion to film not only the release of the munition but also the flight of the munition to the ground and the impact with the target if possible. It took some very hard maneuvering on my part to be in the correct position for the drop. Just as we saw the munition drop from the aircraft, and I was positioning the aircraft to get good photos of the munition in flight, the photographer in the back seat said he dropped the camera. The camera had wedged between the floor and the back-cockpit control stick. As hard as I pulled there was no reaction from the aircraft. We were heading toward the ground at 400 knots and with a seventy-degree dive angle. We had to do something quickly or we would have to eject from the aircraft or hit the ground. The photographer yelled, "Negative Gs! Negative Gs!" His thought was if I pushed the stick forward it may dislodge the camera. That maneuver was against all "logic." Push the stick forward while I was hurtling toward the ground? I am glad I listened to the photographer and pushed the stick forward as far as I could. The negatives Gs came on immediately and the photographer yelled, "The camera is loose, recover!" I immediately pulled back on the stick and recovered at about 2,000 feet above the desert floor. It was a close call but having the right person in the back seat saved my life and a valuable aircraft. This lesson taught me to trust the real expert in any leadership action you take.

Combat Leadership

I was blessed to have a proactive and wise leader during my combat flying time in Southeast Asia. This leader was designated the air liaison officer, or ALO, and had the ultimate responsibility to coordinate all Air Force flying activity with the leader of the Army unit we were supporting in the field. His leadership skills really formed the basis of my leadership philosophy later in my military and civilian career.

The FAC's mission as described in Chapter Two was very challenging and dangerous. What did this ALO do that was so extraordinary?

First of all, this leader was a very competent pilot and flew the mission we were flying as well as any of the other pilots. Second, he fully trusted each one of the young pilots in his unit to do their very best in their flying missions. And third, he would support us to the max if there was ever an issue that was brought to higher headquarters.

One day we were challenged with terrible weather conditions at our forward operating location that were well below our minimums for flying. The Army officer in charge of our unique mission tried to order the Air Force FAC pilots, who were junior in rank to him, to fly despite the marginal weather. We explained that we would not fly because the weather was below our established minimums and this represented an unacceptable risk. His response was he would court martial each one of us if we didn't fly. We held our ground and refused to fly. Our fearless leader went directly to the Army officer and told him that the pilots under his command had the authority to decide when and if the weather was a factor in flying. We never heard another word about flying in weather below minimums again from that Army officer.

On the negative side of this people equation, we had a second ALO who replaced the first one who had gone to bat for us. Based on the incredible leadership of my previous ALO, the replacement officer had a very steep hill to climb to meet the talents of the previous boss. As the only instructor pilot in our unit, it was my responsibility to evaluate and certify the new ALO in our very dangerous combat mission. I was concerned on my first couple of flights with him: he was a terrible pilot. His flying skills resembled a monkey screwing a football. I had a very hard time training him to the point where I was confident that he could safely fly a mission by himself. The location that we flew out of in Quan Loi, South Vietnam was quite remote — just five miles from the Cambodian border — with a short dirt runway of 3,800 feet, no control tower, no navigation aids, and no runway lights. It was quite a challenging environment to fly out of, especially at night. During this ALO's night check flight we had some very serious issues. I believe he scared the shit out of himself, and me, as he was really struggling to control the aircraft, find our target, and safely communicate with the Army troops on the ground we were supporting.

Following this exciting night flight, he announced that we would no longer be flying night missions out of our base. It didn't matter that we had been flying night missions out of our base without navigation aids nor runway lights for the past eighteen months with absolutely no problems. I told him good luck selling that idea to the Army unit that we supported. Obviously, the Army disagreed with his decision, and he eventually relented to let only the young pilots fly the night missions. As a lieutenant I certainly did not mind because I wanted to get as many flying hours as possible.

The actions of these two officers reinforced to me that one of the most important leadership traits a leader can have is competence... the first foundational leadership tenet in this book. I would never step into any leadership role without committing myself to being the very best I could be at that mission. I also was convinced that I would not unilaterally change any procedures or actions unless I knew the full ramifications of making an uncoordinated change at any level.

Leadership Continuation

One important leadership tenet that I observed while I was a base commander in Galena, Alaska was the concept of leadership continuation. We hear stories about this in every war we fight where the leader of a unit is killed in action and then someone steps forward and assumes the lead to carry on with the battle at hand. Of course, in my base commander role I was not killed in action. I was, however, away from the base many times to conduct business at my headquarters at Elmendorf AFB. The Air Force process usually does not allow for individuals to select their assignments nor for organizations to choose who will fill a particular role. Many times organizations are stuck with whoever is assigned to a particular position. This is especially true in the lower ranks of officers and airmen.

At Galena, it was critically important to me to have someone I trusted to carry on the established policies to conduct our important air defense mission. When I first arrived in Alaska the current vice commander, a major, was less than stellar, and I did not trust his judgment or dedication to our mission. For the first four months in my command, I designated my operations officer, another major, to be the acting commander while I was away from the base. I was "saved" when a new vice commander was finally assigned to Galena. This officer was a talented major, and I was duly impressed with the quick action he took to get up to speed on our mission and support programs.

I included him in all the key decision planning and was more than comfortable with his advice during this initial period. I was confident that he could easily step in and run the base while I was away. I gave my vice commander all the power and tools he needed to continue operations in my absence. We consulted together many times to establish this important relationship, and he performed flawlessly each time his leadership was needed. It was quite a relief to know that I could be away from the base and things would run normally until I returned.

The luxury of having a valuable person as a continuity bridge reinforced with me the concept to always have a person, either formally or informally, ready and trained to take my place if necessary. The civilian world provides better options to choose that person, whereas in the military I had to mentor the right person to do that important job. Having the right people in critical positions is a key to a successful organization.

Command Leadership

Near the end of my Air Force career, I had the misfortune to work for a very incompetent senior officer who was a four-star general. We had all heard rumors about his terrible leadership even before he became my boss. As the deputy chief of staff of the organization that this general led I tried my very best to be supportive of him and his leadership style. I tried to convince the staff that the rumors about this general were probably exaggerated and that we would get along just fine. Boy was I wrong. I will talk about loyalty in this situation later in this book. My challenge now was to deal with a very vulgar and not too bright general officer. He would always degrade people in front of the entire staff during his morning staff meetings and most other settings if he didn't like the person's presentation or the person himself. He was always very vulgar in his discussions with other people including other high-level generals, lower ranking staff members, and even high-level international visitors to the command. I had always been taught by previous leaders to "praise in public and discipline in private." This general officer did just the opposite at every opportunity.

A slightly humorous situation occurred early one morning when I got called by the general's aide to get my ass to the general's office. He didn't like one of the staff summary sheets that he had on his desk. He started to really chew may ass... I had never heard such a string of very loud expletives aimed at me in my life. As I left the office five of the other general officers on the staff were waiting outside this four-star general's office. As I walked away the three-star deputy commander said, "Thanks, Tom, for getting him so fired up before we see him."

Many times, he was completely dishonest with his staff and would make up a narrative that most of his immediate staff knew was false. He would wail at his staff officers in the morning staff meetings saying that he had come to the office at 5:00 a.m. to read the classified information before the staff meeting. He then would belittle the staff for not reading the reports before the staff meeting. The immediate office staff knew he never came to the office at 5:00 a.m. and had never read the classified information the before the staff meeting because the safe had never been opened prior to the meeting.

I learned a lot from this general and decided that I would never exhibit any of the leadership traits I observed in him. Even though I was retiring in just one year after his assignment to our command, I carried these "never-do" leadership traits into my civilian and volunteer career.

Leadership Lessons Everywhere

Sometimes leadership lessons can be found in unique places. I had the honor of ministering to an individual through the lay ministry program I volunteered with. The person I was helping actually showed me a leadership trait that I wished I would have had sooner in my career, either my military career or civilian career. He was facing a very serious and difficult legal issue that really gave him no options. As he was trying to save his construction company he got sideways with some very picky employee health benefit laws that really conflicted with paying the salary of his workers. This crisis was brought on by the terrible recession our country was going through at the time. He had a choice: pay for health care or pay the workers' salaries, but he did not have the ability to do both.

My friend was well aware of the potential consequences of his decision but made a judgment on what way to go. One of his employees went to the state labor office and filed a complaint. To make a long story short he was charged, tried, and convicted for not paying his employees' health benefits during the difficult recession. Not one time did I hear him complain about the verdict nor the sentence that was imposed on him. He knew that he had made the correct choice by paying the salaries of his employees and said under the same circumstances he would take the same action again.

During the entire time he was in confinement (work release) he kept a positive attitude, worked extra hard, and got the company back on its feet as the recession began to fade. What this person taught me about leadership was to always do what you think is right, make the decision based on the best information you have at the time, and then deal with the consequences that may follow. He chose not to dwell on the mistake but pressed on to make things better.

People Leadership Lessons

- Know your people.
- Build your leadership skills by watching other leaders.
- Select the best people for key positions.
- Always have a designated "back-up" leader.
- Mentor your back-up leader.
- Learn from bad examples of leadership.
- You will find bad leaders… learn to cope.
- Take responsibility for your decisions.
- Lead the best you can, no matter the circumstances.
- Make changes slowly and purposely.

4

COMMUNICATION

The fourth and final foundational tenet of my leadership model is communication. Communication in leadership is really a two-way street. A leader must communicate to his or her organization and from the organization to the top of the leadership pyramid. Effective two-way communication is vital in any healthy organization whether military or civilian. I have observed that the simpler the communication is the better the organization functions. In the military that keep it simple communication process can be a challenging task. And finally, a leader must be completely visible to the entire organization. Being an effective leader includes the accountability that comes with that leadership position. To know that your followers are watching your every move and following your example keeps you humble and cognizant of your leadership responsibilities. In this chapter I will describe many unique communication scenarios that laid the foundation for what I believe is a healthy leadership model.

Undergraduate Pilot Training (UPT)

As I described in the chapter on competence, I had an instructor pilot (IP) during the T-37 phase of UPT that provided an excellent model of the basic communication skills necessary for an effective leader. As we prepared for each flight, my IP meticulously worked through the most thorough preflight briefing that I had ever experienced in my very short flying career. While the information he provided was very technical, to say the least, this IP keep the communication process so simple that even a twenty-two-year-old lieutenant could understand the concept he was teaching.

He started with the basic review of the mission we would be flying that day, the exact objectives of the flight, the airspace that we had been assigned, what weather we could expect to find, and an exact timeline from getting our flight equipment to walking to the crew bus that would take us to the aircraft. Next, he would cover the exact steps we would take during the aircraft walk around inspection, strap-in procedures, engine start, radio calls, and taxi instructions, including the taxiways we would take and the active runway we would use for takeoff.

He then would go over in exact detail the maneuvers we would be doing, reminding me of techniques that he had showed me on previous flights. For example, he reviewed the exact pressure and movement of the control stick, the specific aircraft gauges I should monitor, and the precise airspeed and vertical velocity indicator readings I should expect on each maneuver. He would then review the procedures, step-by-step, on how to depart the training area, enter the traffic pattern, land, taxi back, and shut down the engine. I felt like I had just completed the flight with all the information he reviewed. But that wasn't the end of the communication process.

Following each flight, he would go over in excruciating detail each and every step of the fight. He critiqued each maneuver, for example saying that on a particular maneuver I was three knots too fast or ten feet too high on another maneuver. But he also reinforced those areas that I performed well on to keep me motivated to get better on each and every flight. I used this preflight and post-flight briefing concept for the remainder of my twenty-six-year flying career. I also tried to teach this communication process to the younger pilots I had the privilege of flying with.

One very telling example of the influence this detailed communication process had on some of the younger pilots occurred during my command of a tactical operations flying squadron in Alaska. I had hoped that the detailed briefing format that I used on my flight briefings had been observed by the other pilots in my squadron. On one particular mission my squadron was going to fly a four-aircraft Dissimilar Air Combat Training (DACT) flight against four F-15's from another squadron in Alaska. DACT simply means aerial combat training by two different kinds of aircraft. My squadron four-ship flight was led by a senior first lieutenant, and the four F-15 pilots were led by their squadron commander, a lieutenant colonel. The lieutenant colonel was not prepared at all for the preflight briefing and missed many key points that should have been covered for a safe flight. While I was one of the four pilots in the flight, I was not the flight lead. The highly competent first lieutenant told the lieutenant colonel that he was going to cancel the DACT mission because the briefing was completely unsatisfactory and could lead to unsafe flying conditions. The lieutenant colonel looked to me and said, "What is this all about?" I answered that the first lieutenant was the flight lead, he was in control of the flight, and ended with, "Your pre-briefing was, in fact, dog shit!" I couldn't praise the young first lieutenant enough for his courage in canceling the flight. I guess my communication skills had been noticed in my squadron.

Combat Communication

I had the good fortune of learning and putting into practice detailed communication skills while I was in pilot training. Little did I know how important that capability would be on my first operational flying assignment. I was tasked as a forward air controller in Southeast Asia as a very young first lieutenant. As I explained in Chapter Two, the FAC's mission was to control every aspect of an airstrike whenever ground troops were involved in the mission. Controlling the airstrike demanded very precise communication skills with many participants in chaotic and rapidly changing environments.

The FAC communicated with the ground forces on an FM radio to make absolutely sure where the friendlies were located and, just as important, where the enemy was located. One slip in the communication process from the FAC to the ground friendlies could be disastrous. Equally important was the communication with the fighters who would drop the bombs. The fighters communicated with the FAC on UHF radios. The FAC had to be absolutely certain that the fighters received the information about the friendlies, the enemy position, the terrain, the kinds of defensive weapons they could expect to see, and a safe exit route if they were hit by ground fire. All of this information was relayed to the fighters before the first bomb was dropped. The FAC would then shoot a white phosphorus rocket at the intended target and ensure the fighters saw the mark. Only then would the FAC "clear" the fighter to drop the bombs. The communication process was continuous. After the first bomb was dropped, the FAC confirmed with the ground friendlies if the bomb had hit in the intended location. Corrections for the next set of bombs would start another round of critical communications between the ground friendlies, the FAC, and the fighters.

The lessons of complete and accurate communication in combat showed me the importance of this skill as a facet of leadership, one that I would experience in the future. You will see examples of this communication technique throughout the remainder of this book.

Flight Test Communications

One of the early missions in my career was in the flight testing. Communication skills were critical to carry out a safe and effective test flight. The aircraft conducting the test and the photo/safety chase aircraft all had to be on the exact page and verse of every step of the flight. The skills I learned from my UPT IP and my combat flying experience were invaluable in the flight test business. The preflight briefing was critical to safely conduct the test flight and meet all the parameters of the test. Some of the items covered were the takeoff time, range airspace, expected range weather, operating altitudes, minimum drop/launch altitudes, expected flight path of the munition, exact parameters that needed to be filmed, potential fragmentation pattern of the munition and or target, emergency procedures, and recovery/return to base options. Each planning factor was critical and had to be covered in fine detail for each mission.

Just as important was the post-test flight debriefing. Did the munition drop at the correct speed and altitude? Were there any anomalies to the munition when dropped? Did the safety/photo chase pilot observe any anomalies? Did the munition hit the target? All of these observations following the flight were factored in for subsequent follow-on testing. The success of any test flight really depended on the preparation for the flight and the candid debriefing following the flight on all the desired test parameters. This detailed communication process was a foundational leadership idea for many different areas.

International Communication

Miscommunication in an international setting can cause some issues. I had a very intense situation occur while I was instructing at the U.S. Air Force Air Command and Staff College. We normally had about eighty international military officers from countries around the world attend the ten-month ACSC course. One of my students, a major from the Saudi Arabian Air Force, brought his family to the school and it was a real pleasure to get to know them and to hear his many stories of his time in the military. This major and his family loved to eat lamb for many of their main meals; however, he had trouble finding fresh lamb in Alabama. Being a creative person, he found a farmer on the outskirts of Montgomery who sold him two live lambs. He put the lambs in his car, drove them to his apartment in Montgomery, and then proceeded to slaughter them on his patio. Obviously, his U.S. neighbors in the apartment building were none too keen on this activity and called the police.

The local sheriff arrived on scene and called the ACSC hot line to describe the problem. I was listed as the faculty contact for this major, so I was contacted to go to the apartment as there was a real problem going on. When I arrived on scene there were many neighbors standing around the Saudi major's blood-filled patio. The local sheriff briefed me on what had happened with the two lambs. I asked him if there was a law against slaughtering animals on patios in Montgomery. He said that there was not, in fact, a law against this activity. I explained the cultural differences about preparing fresh meals at home and told the sheriff I would discuss the perceived problem with the major. I then explained the cultural differences with the major and asked him in the future not to do any slaughtering on the patio. He was a bit confused but said he understood, explaining that this was a very normal practice in his country. This situation just confirmed that in any effective communication process there may be cultural differences that cloud the communication process. Fortunately for the major, he found a source for fresh, already slaughtered lamb in Atlanta.

Unique International Communication

During my tenure as a faculty instructor at ACSC I was privileged to experience another interesting international communication scenario. I am certain that this scenario would be difficult to duplicate if it were not for the unique international setting at ACSC. An Israeli Air Force major, an F-4 weapons system operator and veteran of the 1982 Arab Israeli War, was assigned to my student seminar; each seminar at ACS was comprised of at least ten U.S. officers and two international officers.

As part of our curriculum, we studied many of the famous air battles fought throughout the history of warfare. One of the most recent air battles was the 1982 conflict in Lebanon. My Israeli student mentioned to me that he had flown in that campaign and had shot down a Syrian Mig-21 during the battle. What a real-life story he could share about the actual air battle. As fate would have it there was a Syrian officer attending ACSC that year who had been a pilot on one of the nearly one hundred fighter aircraft shot down by the Israeli pilots. This presented a unique opportunity if the two were willing to tell their war stories. So, in close coordination with the faculty instructor of the Syrian pilot, we came up with a plan for both of these officers to tell their side of the shoot down story. We combined the two seminar groups for the very intense discussions about aerial combat.

I was not sure how the discussion would progress, but each officer was very respectful of the other and recalled in minute detail what they were doing on their specific missions. They provided great insight into how their two countries conducted aerial combat. The coordination process to bring these two individuals together — whose countries faced each other in recent combat — for the benefit of the entire class is another example of how important effective communication can be at any level of leadership. These two professional officers respected each other and realized they were both just doing their assigned missions. Positive communication helped the students from both seminar groups get a very personal inside look at aerial combat from multiple perspectives.

More International Communication

As I mentioned in the mission chapter of this book, I was selected to teach for one week at the British Staff College in Bracknell, England.

Three other faculty instructors from the Air Force Command and Staff College (ACSC) flew with our spouses to England to present the most current information about the U.S. Air Force. I was scheduled to present a one-hour session on future Air Force programs. The commandant of ACSC, a brigadier general, also accompanied the faculty instructors for the week. When we arrived at Bracknell our commandant told us to try to incorporate some local culture into our briefings so we could better connect with the British students.

The night we arrived in England my host, a British wing commander, and his wife took us to a local pub for some merriment and fun. What a great time we had! I did notice a special food item listed on the pub menu and felt that I could use the name of the food special in my briefing. On the third day of teaching, I was scheduled to present my briefing to the class of about 300 students. I told the audience that I normally don't recommend to my students at ACSC to apologize about the briefing they were about to give. But in this situation, I just had to offer my apologies ahead of time if I offended anyone because of the difference between U.S. English and British English.

You see what I saw at the pub a few nights previous concerned me as the food special for that evening was faggot balls. I certainly knew that term had meaning in the U.S., but I thought maybe England had also come a long way in their social engineering process and that it was just a known term. I got a very loud and boisterous response from the audience when I mentioned the faggot balls special I saw at the pub. It seems that in England, faggot balls is another term for meat balls, and faggots are made from offal, usually pork, and from bits of the animal that are generally discarded (the heart, liver, and intestine). My reference to the faggot balls was such a hit that at the formal dinner two nights later, I was presented with a container of frozen faggot balls by the commandant of the British War College, a two-star general, as a token of my acknowledgment of the English culture. This humorous situation gave me another key data point in making sure you understand the communication nuances of different cultures.

Command Communication

Being a commander or the ultimate leader of an organization requires some special communication abilities. During my two command assignments I used several communication techniques that I had observed from other leaders during my career. While each command was somewhat different, the basic concepts that I used were the same.

While in command of my flying squadron in Alaska I used a technique that I had seen applied very effectively when I was a line pilot flying many different types of aircraft. On my first day in command, I summoned the entire squadron, all pilots and support staff, to a Commander's Call. In the Commander's Call I explained my detailed leadership philosophy and expectations to every member of the squadron. I outlined my expectation that everyone should complete their assigned tasks to the best of their abilities. I told the squadron I was not afraid to make a decision about anything based on the information I had at the time of the decision but reiterated that if anyone had better or more current information regarding my decision, I would gladly review that decision and change my mind if necessary.

Another important item I impressed on the squadron was the absolute importance of flight safety. I told the entire squadron that anyone in the chain of command had the authority to raise the "red flag" and stop operations if something didn't look safe. That meant the two-stripe admin technician, the life support expert, the pilots, and any person regardless of rank could call a halt to an activity. Safety was paramount. We did great in this area and had no safety incidents during the entire time I led the squadron. We were so good that during an Alaska Air Command Standardization and Evaluation (STANEVAL) Inspection our squadron received an "Outstanding" rating... the highest rating possible for our adherence to standards and safety awareness.

When I became the base commander of a remote air defense alert site I used much of the same direct communication with the members of that squadron. One communication technique I learned from my time as the T-33 commander was to eat with the squadron airmen in the dining hall whenever I was there. So, I made a promise that every day, other than during the Tuesday officers' staff luncheon, I would sit with random members of the squadron at breakfast and lunch to get a feel for how things were going. My first attempt was kind of funny because I sat with two very young supply airmen and asked them how their assignment was going. I couldn't get a peep out of them. I tried getting them to talk to me, but they had never seen their previous commander sit with the airmen, and they didn't know how to act or what to say. I believe they were afraid to say anything. But I was persistent and finally after the third day when I sat with a group of firedogs (fire department personnel) they opened up a little bit about how things were going. They did not like the music at the club or the food at the bowling alley, and the one thing they really did not like was the very small bowls for the ice-cream machine in the dining hall. So, I told them I would fix that and had them follow me. I took them up to the ice cream machine, grabbed a large water glass, and filled it with soft serve ice cream. They said they weren't allowed to do that. I told them starting today they could use the glasses until we got larger bowls. Following lunch, I immediately stopped by the master sergeant in charge of the dining hall and told him to get larger servings bowls for the ice cream and to plan on making as much ice cream each day as his supplies allowed. Problem fixed!

Another communication technique I learned from a former commander was to "get out and see the airmen" to learn what they do a on a day-to-day basis. I unofficially called it "leadership by wandering around." Whenever my schedule would allow it, I only missed about four weeks during the year, I visited a single department or division each week and spent the entire morning with the airmen. I didn't miss a single location. I did have some very interesting experiences with the airmen. While doing my half day with the fire department I got to drive one of the huge fire trucks, suited up in the firefighting equipment, complete with air tanks, and simulated a fire rescue with artificial smoke and all. It was quite hard work for a forty-one-year-old lieutenant colonel. But the airmen were great and made allowance for the boss. I worked all morning in the supply warehouse sorting aircraft parts for inventory control. I was so surprised to find hundreds of parts for an aircraft, the F-106, that had not been stationed at Galena since the early eighties. My favorite location to visit was the dining hall where I would "help" the folks prepare the meals for the day. They always were concerned and said, "Here comes the colonel, he will change the recipe again." I also made it a point to never be off base during any holiday, and I always worked the food service line handing out food to the airmen on those special days away from family. Many folks were very surprised to see me behind the serving line. I know how tough it was to away from family during the holidays, and it was the least I could do for the squadron members.

Another communication technique that I implemented at Galena was the Commander's Gram program. This program gave every person assigned to Galena, military and civilian, an opportunity to communicate directly with me on any issue that they had. I had a twelve-inch-by-twelve-inch box with a narrow slot at the top and a padlock on the cover installed in a prominent location in the base post office. I had a stack 4.0-inch-by-5.5-inch forms attached to the box. The form was very simple: "What do you want the commander to know or answer?" Including a name was optional, but if I didn't have a name on the form for a question, I may not be able to get the answer directly to them. Everyone assigned to Galena visited the post office at least once a day to receive mail and care packages from home. I was very curious about what kind of response I would receive as I had mentioned the program at my first Commander's Call at Galena and told them names were optional on the forms. Well, it didn't take long for my first Commander's Gram to appear as I checked the box at least twice a week.

The first Commander's Gram was a two-page list from someone who identified each and every sexual relationship that was supposedly taking place on the base. Not only did I get names, but this person also listed room numbers and if the people involved were married. Wow — what a hot topic to deal with! I called the chaplain to my office to discuss this issue, and we really didn't have an answer of what to do. We were both curious who the source of the information was and if that person was jealous of not being in a relationship. These relationships would certainly raise some leadership issues for me later in the year.

The Commander's Gram program seemed like a success as I usually received ten to twelve submissions each month covering a wide range of topics from promotion chances to requests for reassignments to specific mission support requests. I do believe that through my daily meals with the airmen, the weekly shop visits, and the Commander's Gram program I was able answer many of the questions that were being asked by the squadron personnel, and they started to feel more comfortable talking to me during the meals.

These short stories are prime examples of the importance of two-way communication in any organization.

Immediate Communication and Feedback

I also made it a policy to in-brief every single military and civilian member who was newly assigned to our squadron. I broke the in-briefing down into two groups: the officers and the senior noncommissioned officers (E-7, E-8, E-9) and then the junior enlisted personnel (E-1 through E-6). I was very straightforward with the information that I passed along. I first reminded them that even though they were assigned to a remote Air Force base we would follow all the rules and standards just like any base in the lower forty-eight states.

I told them that I expected each one to do their job to the best of their ability and also let them know that I had a "open door" policy if there were ever any issues affecting their ability to get their job done. I explained the importance of our air defense alert mission and reminded them how very important each part of their mission was to the overall success of the squadron. I reminded then that it was tough being away from family for one year and that we would do everything we could to mitigate any problems.

Another important time of communication was when an individual, either military or civilian, transferred from the squadron to another base. Before these departures I required them do an individual out-briefing with me so I could correct or enhance any leadership issues that may have come up during the year. I learned a lot from these out-briefs and, where appropriate, tried to implement the recommended changes. This process really emphasized the importance of two-way communication in any organization.

Headquarters Communication

Following my assignment to Galena I completed three different staff positions at the headquarters of North American Aerospace Defense Command (NORAD). Each position required the same and sometimes expanded requirement to effectively communicate.

My first headquarters experience was as the chief of NORAD's counter-narcotics planning division. At the time (late eighties and early nineties) the entire country was struggling with a plan to intercept the vast quantity of illegal drugs that were flowing from South and Central America into the United States and Canada. NORAD's role was to conduct air interdiction actions of any illegal drug trafficking flown into the U.S. or Canada. When I arrived at NORAD in October 1989 the counter-narcotics plan was pretty well written but not yet embraced by all the various parties involved in the drug interdiction program in the U.S. and Canada. My task, as the chief of the counter-narcotics planning division, was to "sell" and coordinate the NORAD plan with all the various players involved. Communication skills were vital in this effort.

In order to get the NORAD plan accepted at every agency I decided that we needed to put together a traveling "dog and pony show" to explain each facet of the NORAD plan. This required detailed planning and an exact communication plan, specific to each location, to succeed. Our targets were varied, spread throughout the U.S. and Canada, and involved many agencies outside the military. We had to tailor each briefing to the specific agency we would present the plan to. The agencies included the Air Staff and the Joint Chiefs of Staff at the Pentagon, the Office of National Drug Control Policy (ONDCP), the Drug Enforcement Agency (DEA), the Federal Bureau of Investigations (FBI), and U.S. Customs and Border Patrol in Washington, D.C., the Canadian NORAD Region headquarters in North Bay, Ontario, the Alaska NORAD Region in Alaska, and the Continental NORAD Region in Florida. In addition, there were several lower-level agencies that we briefed as they would be doing the actual interface with the NORAD plan. As you can imagine, we had many different communication presentations that we tailored to each agency. The bottom-line guidance on the communications presentations was to keep the information as simple as possible and be prepared to honestly answer any questions that may arise. I also directed that we capture all the questions that came up in previous briefings and include the answers into the next briefing that we presented. I believe that our hard work "selling" the NORAD plan was effective because we had many successful intercepts of illegal drug flights into the U.S. and Canada. However, the drug cartels found new ways to smuggle the drugs int the U.S. as we had a very porous border along the international boundary with Mexico.

More Headquarters Communication

After one year as the counter-narcotics division chief I was promoted to colonel and assigned as the deputy director for the NORAD planning division. Just because someone wears the rank of colonel, the communication process does not stop. My first task in the new position was to do a study of all the various NORAD locations around the U.S. and Canada that had fighters on five-minute alert status to intercept any Soviet aircraft or drug smuggling aircraft that would try to penetrate our sovereign airspace. My initial approach was to do a very scientific analysis of the situation from radar surveillance capabilities, current alert site locations, and aircraft capabilities of the U.S. and Canada alert fleets. The NORAD civilian, a math expert with a PhD, who put the study together presented me with the results. While the numbers and analysis all made sense, the format of the briefing sucked. Readers really had to pay attention to the math and some very basic leaps of logic to come to the final conclusion and recommendations of the study. While I could understand the logic and the math involved in the analysis, and I agreed with the final recommendations in the study, I knew we could never "sell" the results with the current briefing format. I directed my team to simplify the briefing so that even a fighter pilot could understand the logic of how we arrived at the final recommendations. Once again, the concept of keeping it simple was in play.

After the briefing was revised and simplified to my satisfaction, it was time to brief the results to the NORAD commanders (eight general offices from the U.S. and Canadian) who were attending the annual NORAD Commanders' Conference. On the very icy winter morning of the Commander's Conference, the lieutenant colonel scheduled to present the Alert Study briefing called twenty minutes before the start time to explain that he had been in a multiple-car accident and could not present the briefings. Wow, I was sure glad that I had been deeply involved in the Alert Study process and the final briefing product as I became the designated briefer. The presentation went off without a hitch but, as expected, the generals on the staff wanted this Alert Site study briefed at various locations around the U.S. and Canada.

So, as we did with the NORAD counter-narcotics plan we prepared to bring the Alert Study findings to the field for "buy in" from the various organizations involved in the Air Defense mission in the

U.S. and Canada. The organizations affected by the study results and recommendations were Headquarters U.S. Air Force, The National Guard Bureau, National Defence Headquarters in Ottawa, the Continental NORAD Region in Florida, the Canadian NORAD Region in Canada, and the Alaska NORAD Region. Since I had been at the very pointy end of the combat stick in Alaska, I knew the headquarters in Alaska would appreciate getting the in-person briefing on the NORAD Alert Site study. These actions reinforced my belief that critical and timely sharing of information in the communications leadership role is essential to get things done.

As I progressed in my staff duties, I was promoted to deputy chief of staff of NORAD. I worked for both a Canadian three-star general and a U.S. Air Force four-star general. My main task was to ensure that the entire NORAD staff worked efficiently and provided the two generals with accurate and timely information. My initial and most time-consuming daily task was to approve all paperwork that would be presented to the two generals. My approach was to apply a "can a fighter pilot understand this information" review on any product that was going to the generals. Both generals had been fighter pilots earlier in their careers.

Going forward, I used the analytical approach I had learned in my accident investigation days to review the sometimes highly technical information provided to the generals. Whenever I had a question about any of the material going forward to the generals, I had the author, or as we called them the action officer, come to my office and explain the point he was trying to convey. My main goal was to communicate, in the most logical and straight-forward fashion, what a particular piece of information meant. After about two months of some very intense questioning of the material coming across my desk, I believe the staff finally got the clue that a simpler message had better chance of surviving my scrutiny to make it to the generals' desks. Even at the highest level of command the importance of effective communication is very important.

Cooperative Communications

Since NORAD had such a diverse mission and the four-star general in command of NORAD was also the commander of U.S.

Space Command, the importance of effective communication and coordination between the two staff elements was critical. I had close coordination with the U.S. Space Command deputy chief of staff, a U.S. Army colonel. Together we came up with an effective communication coordination process. Every two months we would invite all of the deputy directors from the NORAD and USSPACECOM staff to meet for lunch at the officer's club. These were all either Canadian or U.S. colonels and a couple of one-star generals. Our main goal of the luncheon was to make sure the communication process among the staff elements was effective. We also left time for the attendees to describe any coordination issues they were having with the communication process between the two commands. This very simple coordination meeting paid huge dividends to both commands. Effective communication was alive at the highest levels, and the hope was that this effective communication process would filter down to the entire staff elements of both commands. It certainly made my job easier when I was trying to get programs coordinated and approved by the respective staffs and both generals.

Cross-Cultural Communication

I experienced a very unusual cultural communication issue when I was the deputy chief of staff of NORAD. In the mid-nineties the U.S. and the Soviet Union were starting to become "friends" in a convoluted sort of way. The two countries hosted many cross-military visits to start the trust process. One important visit from the Soviet Union to the U.S. was by the Soviet-equivalent of their Secretary of Defense to our Cheyenne Mountain Complex located at Cheyenne Mountain Air Force Station (CMAFS), a short distance from NORAD at Peterson Air Force Base in Colorado Springs, Colorado. At the height of the cold war in the late 1950s, the idea of a hardened command and control center was conceptualized as a defense against long-range Soviet bombers. The Army Corps of Engineers supervised the excavation of Cheyenne Mountain and the construction of an operational center within the granite mountain. The Cheyenne Mountain facility became fully operational as the NORAD Combat Operations Center on February 6, 1967. Under what became known as the Cheyenne Mountain Operations Center (CMOC), several units supported the NORAD missions of aerospace warning, aerospace control, and maritime warning for North America. This location was one of the most important facilities for the defense of the U.S. and Canada. To have our number one adversary visit the Cheyenne Mountain complex was a huge deal with much interest from not only our military organizations but also Congress. So, you can imagine the planning that went into the visit. We covered almost every contingency that could possibly come up to make sure this visit went as smoothly as possible.

We prepared comprehensive plans for everything from the location the Soviet delegation would be staying at to the normal interface we would have with our military leadership and also the leadership of the city of Colorado Springs. We had a solid transportation plan and previewed each briefing the units at Cheyenne Mountain would present to the Soviets. We also had a Russian translator assigned to the briefing team in case there was a language issue during the briefings. We were all set and ready to go when disaster struck. During the very first briefing in Cheyenne Mountain, the Soviet Secretary of Defense mentioned that he could not understand a word of what our translator was saying. We were really confused until one of the members of the Soviet delegation informed us that our translator was speaking a Russian dialect that was not recognized as official Russian. We were informed that the Russian language has three very distinct dialects, and the principal member of the visiting Soviet delegation was fluent in one dialect, not the one our translator was using. We were in quite a bind. Fortunately, one of our Soviet intelligence experts from the NORAD staff was sitting in the back of the room during the briefing and volunteered to be our translator because he was fluent in the Russian dialect that the principal understood. Our bacon was saved, the remainder of the briefings went without a glitch, and we successfully hosted the Soviets at our facility.

The main lesson that I learned from this episode was that there are many small but important pieces of any communication process, and it is best if you are aware and plan for any issues that may arise, especially when dealing with cross-cultural scenarios. You can be assured that all future NORAD visits from non-English speaking delegations were scrutinized, and the dialects of the visitors were confirmed well before the visit began.

Translate This!

While I was the division chief of the NORAD counter-narcotics planning division, I had many opportunities to present the NORAD counter-narcotics briefing to many organizations all the way from the U.S. Joint Chiefs of Staff to the French Air War College. A contingent of French students from Ecole de Guerre were planning a visit to the NORAD headquarters as part of their cross-military education process. In addition to the formal NORAD command briefing, it was decided to present the visiting college members an update on the NORAD counter-narcotics plan. I was selected to present the briefing and would have to work with a real time, English-to-French interpreter while I presented the briefing. Because I have a French-sounding last name I asked one of our Canadian staff officers who worked for me to prepare a greeting in French so I could formally welcome the visitors before I presented the counter-narcotics briefing. Because of the high level of interest in this visit, we did a complete dry run of all the briefings that would be presented a couple of days before the actual visit.

I really practiced the welcome phrase that I was going to use and was confident I would not screw it up. During the practice run of the briefing I confidently opened with, *"Ma merde de chat dans ta valise."* The remainder of the briefing was completed with no problem ,although it was fairly slow waiting for the simultaneous conversion from English to French. Following the dry run briefing I asked if there were any comments on the counter-narcotics briefing. A young French-Canadian Captain in the back of the room asked me if I knew what I had said in French. I said, "Of course, I welcomed the group to the briefing." He told me that what I actually said was: "Your cat shit in my suitcase." The Canadian major who prepared my welcome in French was doubled over in the briefing room laughing so hard about what almost happened. Needless to say, I stayed with an English welcome and the briefing to the French War College went off just fine.

Defense Contract Communication

Following my retirement from the Air Force I became a consultant for a Fortune 500 company that dealt with satellites, satellite control systems, the intercontinental ballistic missile (ICBM) program, and very high-level space and space systems research and development programs. The possibility of miscommunication with all of these technical programs was very high. As a consultant for this company, I was the gate keeper of the kind of information that was communicated to our main customer, the federal government and, more specifically, Air Force Space Command, the Air Force Research Laboratory, and NORAD. Using the keep it simple approach, I had to convince the very smart PhDs who worked for my client how to present technical material that could be understood by the non-technical decision makers in the government. Leveraging some of the coordination and communication techniques I used as the deputy chief of staff helped me solve many potential communications issues before they became problems.

The first priority was to review every document that the company was going to present to the government. This included written as well as the visual material they would present in person. Many times, I was able to point out that some of the material was too technical and confusing, and that it did not lend itself to a self-evident conclusion. After all, we were trying to sell a product or a concept to a customer.

I was very aware of the individual peculiarities of various general officers and their likes and dislikes of the way material was presented to them. These were the people who would make the decision on whether they would purchase the product or concept from my company. A prime example of this approach involved a two-star general in Air Force Space Command who always wanted the bottom line of any presentation to be the second slide shown to him. I learned of this preference through regular contact with the general's staff who would tell me what the general liked and did not like in presentations given to him.

Another general preferred diagrams of the material being presented, another general liked to have a hard copy of the presentation delivered to his office twenty-four hours in advance of the presentation, a different general wanted a hard copy of the presentation in his hand so he could write notes to himself, and yet another general would lose interest in the presentation if there were more than ten slides. While this certainly caused problems for my client, the fact that they knew in advance of the likes and dislikes of the decision makers they were presenting the material to made for some very successful presentations and, better yet, signed contracts.

Volunteer Communications

Immediately after my Air Force retirement I became involved as the director of a Christian-based volunteer lay ministry program at my local church. The primary goal of the ministry is to "walk" with people who are going through difficult times in their lives. We don't try to solve their problems, rather the volunteers in this program are only there to listen to their concerns about their problems. We let the individual come to the solution themselves. The volunteers are prepared through fifty hours of intense training followed by continuing education classes and peer supervision twice a month.

One of the most important ministry lessons we teach is the art of listening. Listening is the foundational communication skill we require of our volunteer lay ministers. During training the ministers get a chance to review and practice all facets of the listening skill, including how to position yourself, how to use reflective listening skills, and, most importantly, when to be quiet. These skills are critical in helping the person we are caring for discover their own solutions. Individuals might not be able to discover their own solutions to problems if the lay minister is talking more than listening. If a person comes up with their own solution, they tend to own that solution and act to implement it. As I look back on my career in the Air Force and civilian life, I can certainly see the benefit of effective listening; I wish I would have learned those listening skills earlier in my career. Even though I was done with my Air Force career, I used the listening techniques in my civilian position every chance I could.

COVID-19 Communication

One of the most in-my-face leadership lessons happened during the COVID-19 pandemic. The Christian ministry program that I led was based almost entirely on effective and in-person communication between the lay minister and the care receiver. Once the lockdowns from the COVID pandemic were put in place, our program volunteers had to rely entirely on either phone or text communication between each other and with their care receivers. This certainly was not the most effective way to help and support care receivers during difficult times in their lives. But because of the foresight of our ministry leadership team, we trained our lay ministers how to use either Zoom or Microsoft Teams platforms to help them communicate with their care receivers in a more personal way than text or telephone. We also implemented a Zoom or MS Teams program for the leadership team to conduct their monthly meetings. Our program continued to thrive despite the pandemic because we realized that sometimes technology can help in the very important communications factor no matter what organization you are supporting. An organization will cease to function if all communications are lost.

Communication Leadership Lessons

Keep it simple, stupid (KISS).

- Emphasize details.
- Coordinate for effective communication.
- Communicate more rather than less.
- Be an effective listener.
- Learn to "sell" your message.
- Encourage two-way communication... person-to-person is best.
- Recognize cultural differences.
- Be courteous with your adversary.
- Be aware that technology can assist in the communication process.

5

STANDARDS

Once the basic foundation of a leadership model that includes Competence, Mission, People, and Communications has been built the next step is to start adding the walls to the leadership model. An important structural leadership element is the establishment and insistence on standards. Standards are simply the defined criteria for success of any organization. I believe an effective leader must unapologetically set high standards and then have the courage to enforce those standards. The stories in this chapter below will give you an indication of the kinds and scope of standards that contributed to building my leadership model.

Early Standards

As outlined in my first chapter, my ROTC experience at the University of Portland in the mid-sixties established the fact that standards were important to the health of any organization. The ROTC program had a minimum GPA standard — 2.5 — to progress into the advanced ROTC curriculum for the junior and senior years. Because I was goofing around so much during my first three semesters, I had to really hump to get my GPA up to the required standard to continue in ROTC. This lesson showed me that standards are important as they set the minimum acceptable level for any activity. Without standards there is no way to effectively measure the success or failure of an organization. This initial lesson in establishing and enforcing standards was observed in every single mission I was assigned to or volunteered for in my future careers.

Flying Standards

It would be almost self-evident that there must be standards in place in any flying activity. There are Air Force standards, Federal Aviation Administration (FAA) standards, and standards defined by the flying mission. Fortunately, as I described in the first chapter, my T-37 instructor pilot provided a perfect example of how to meet the Air Force standards. Each maneuver we performed — from taxi to take off, entering the training airspace, flying a specific maneuver, entering the traffic pattern, landing, and the taxi back — was stressed on each and every flight.

There was a very funny scenario that played out during the T-37 phase of my undergraduate pilot training (UPT). The T-37 was a side-by-side seating aircraft with the student pilot in the left seat and the instructor pilot in the right seat. To get an idea of the importance of standards, one of the T-37 instructor pilots flew with a small red plastic hatchet on every flight he instructed. If the student pilot made a mistake and did not meet one of the many standards of flight, he would hit the student on the helmet with the hatchet. At the time, the standard Air Force flight helmet we wore was white. It would be very obvious to any observer after the flight just how the student pilot performed by the number of red marks on the helmet. I did fly with this instructor one time and had my share of red marks on my helmet. The goal when flying with this instructor was to see how few red marks we would receive on any one flight. This is kind of a corny example, but the hatchet certainly gave us an idea of the importance of flying standards.

In the flying world there is no margin for error. The standards at UPT were in concrete and any deviation from those standards could result in a very bad outcome. If a pilot was directed to fly at 250 knots below 10,000 feet, that pilot had to be at those standards. If the calculated approach speed was 140 knots, the aircraft had to be flown at 140 knots. If we were told to fly on a certain radial for an instrument approach we had to be on that exact radial or be in violation of the airspace we were flying in. What I am trying to say is that during flight there are many standards that must be followed, and my flight training was a good example of why standards are so important. My flying experience established a very strong basis for following and enforcing standards in my future Air Force and civilian careers.

Combat Standards

If any time a leader must establish and follow standards it is during combat operations. My experience with standards during combat come from my 535 combat sorties that I flew as forward air controller (FAC) in Southeast Asia in 1970-1971. While I was just a twenty-four-year-old lieutenant, I certainly learned the importance of setting and following standards. In fact, one of the most enduring set of standards was taught to me while I attended the Air Force Survival School in Washington State before deploying to the war.

During a four-day period in this survival school we were exposed to what it would be like if we were held as prisoners of war (POW), complete with intense interrogation, confinement in a prison-like environment, and generally tough treatment of the students. The concept of this training was to teach us how to survive and resist the enemy in the event of being captured.

Service members learn the Military Code of Conduct when they begin their service, and this code was certainly reinforced during the survival training that I endured. The six basic tenets (or standards) are:

1. I am an American, fighting in the forces which guard my country and our way of life. I am prepared to give my life in their defense.
2. I will never surrender of my own free will. If in command, I will never surrender the members of my command while they still have the means to resist.
3. If I am captured, I will continue to resist by all means available. I will make every effort to escape and to aid others to escape. I will accept neither parole nor special favors from the enemy.
4. If I become a prisoner of war, I will keep faith with my fellow prisoners. I will give no information nor take part in any action which might be harmful to my comrades. If I am senior, I will take command. If not, I will obey the

lawful orders of those appointed over me and will back them up in every way.

5.   When questioned, should I become a prisoner of war, I am required to give **name**, **rank**, **service number**, and **date of birth**. I will evade answering further questions to the utmost of my ability; I will make no oral or written statements disloyal to my country and its allies or harmful to their cause.

6.   I will never forget that I am an American, fighting for freedom, responsible for my actions and dedicated to the principles which made my country free. I will trust in my God and in the United States of America.

These standards were so ingrained in me that to this day, nearly fifty years later, I can still recite each article of the Code of Conduct.

In addition, in combat we were also "restrained" by The Law of Armed Conflict (LAC) and a specific set of rules of engagement (ROE) that we had to follow. I did get a short introductory LAC briefing upon my arrival in Vietnam. I learned the specific ROE with on-the-job training as I was checked out as a combat-ready FAC. The basic premise of the LAC was to regulate the conduct of armed forces to protect the fundamental human rights of those involved in the conflict. The ROE directives were meant to describe the exact circumstances under which I could employ any of the fire power available to me as a FAC.

Rules of Engagement (ROE) were directive in nature and compliance was required by all U.S. military forces carrying out activities in the war. Supplementing these rules, and usually more restrictive, were the operating rules and policies established by the commander in a specific area; in my case it was the commander of 7th Air Force. Rules of Engagement formally stated what was permitted or forbidden in air operations. These ROE came directly to me from 7th Air Force; however, the ROE was usually specifically outlined and preliminarily approved by the secretary of defense with ultimate approval by the president. The ROE delineated when, where, how, and against whom military force could be used and the implications if we didn't follow them. As a young lieutenant I wondered how we could ever win this war with such severe restrictions. But I soon realized that these parameters or standards were necessary if we were to avoid a real free-for-all war.

Combat Flying Standards

In addition to the described standards for a POW, the Law of Armed Conflict and the Rules of Engagement, there were also the obvious standards of just flying the airplane within establish parameters and controlling an air strike. We had some minimum altitude requirements that we were to obey unless the expediency of the situation dictated otherwise. We had specific standards on how close we could drop munitions near our supported ground forces, and we had specific protocols of how we would brief the various fighters we controlled on exactly how we wanted the munitions to be employed. Flying combat missions was really constrained by many different standards from where, how, and when we could engage the enemy, but I learned a valuable lesson flying combat: there will always be standards to meet and enforce no matter the mission.

Test Standards

I described in some detail the flight test mission that I flew at Eglin AFB in Chapter Two. Standards were absolutely critical in any test flight we conducted. Some of our complex and sophisticated munition test missions required us to fly precisely at the preplanned altitude, the exact preplanned airspeed, and the specific dive angle or launch parameters dictated by the test plan. If any of those tests flight parameters were not met the entire test could be invalidated and that could cause a huge loss of resources for future testing. Even the photo/safety chase missions had predetermined flight parameters or standards that had to be accomplished to document the test flight of the munition.

Standards were also everywhere in the other test missions that I supported. I was designated as a certified range safety officer for some CIM-10 Bomarc missile test launches. The Bomarc was a supersonic (2.5 time the speed of sound) surface-to-air missile that was used for air defense during the cold war. By the time I got involved in the test support of the Bomarc, the missile was being used as a target for air-to-air live missile firing tests. My position as the range safety officer was to monitor the engine start and lift-off of the Bomarc to ensure a safe flight trajectory out over the test range in the Gulf of Mexico.

For the Bomarc launches, I sat in a concrete bunker about fifty yards from the launch site that was right on the beach along the main highway running east and west from Fort Walton Beach to Pensacola, Florida. I had a very sophisticated tool to determine the correct launch parameters of the missile: a thirty-six-inch long, one-quarter-inch thick steel cable that was stretched vertically between two points on my bunker. If the missile trajectory was to the left of the cable, the flight was deemed safe, but if the trajectory drifted to the right of the cable it was unsafe and could impact on the land side of the test range. I would call out the trajectory of the missile and if all was well with the trajectory, I would say nothing; if, however, the trajectory was to the right of my cable by even a quarter of an inch I would call out, "Right, right, right!" By the third call of "right" another range safety officer would activate the self-destruct system on the missile, and it would be destroyed before it could impact civilian property. The standards for the safety of this mission were critical. If we didn't follow exactly the standards for a safe flight of the missile there could have been catastrophic results. I will relate another example of the importance of standards in the test business later in this chapter.

Contract Standards

Following my time at Eglin AFB I was assigned to support the PQM-102 unmanned drone program at Holloman AFB. The PQM-102 was a modified F-102 supersonic, single-seat fighter cable of flying unmanned for live air-to-air missile tests. I had two duties in this program: first, as assistant operations officer to ensure all operations of the drone program were carried out successfully and safely, and second, as a government flight representative (GFR) and a quality assurance evaluator (QAE). My task in the second role was to ensure that the civilian company that was awarded the PQM-102 contract meet all the requirements designated in the contract.

The PQM-102 contact was very specific and laid out in detail all the standards the contractor had to meet to satisfactorily perform the drone mission and, hence, get paid for their effort. Some of those standards included having an operational drone available to fly at least three times a month, maintaining a mission ready status of the drones at 85 percent, keeping on hand a necessary level of support equipment and supplies for the drone, and employing the necessary drone operators to support the mission. My job was a continual check of the various standards outlined in the contract, and this gave me some insight in to how important it was to have the standards of any organization fully understood by the members of that unit. This knowledge would be invaluable to me when I moved on in my career to levels of greater responsibility.

Command Standards

Anytime you are in command of an organization one of the primary goals is to ensure that the various standards, for not only the mission but also for the unit, are maintained by every member of the unit. For a commander to do anything else is a clear dereliction of duty. An important part of the standards process was to regularly communicate the individual and mission performance expectations to every member of the unit. One slip opened the door for more noncompliance with the standards that had been set. The short vignettes in this section highlight some of the many different standards that were expected to be met during my career.

Flying Standards II

As the commander of the 5021st Tactical Operations Squadron I was keenly aware of the necessary standards that had to be followed for safe flying operations. One of the techniques I learned early in my flying career was to incorporate a formal review of various emergency procedures during each preflight briefing in my squadron. One of the pilots in the group briefing would review the exact procedures in case of an emergency during the upcoming flight. I also instituted an emergency knowledge exam that was required each quarter by every pilot flying in my squadron. That included not only me as the commander, but also the colonels and other headquarters staff who flew in our squadron and every pilot assigned to the squadron. This attention to the emergency review standards paid off for the squadron as we were awarded an outstanding rating, the highest score possible, during our annual standardization and evaluation inspection by the headquarters of Alaskan Air Command.

Military Standards

The military obviously had many personal and mission standards that every member of the unit was expected to follow. These standards ranged from exactly how to wear the uniform to personal appearance, greetings, maintaining body weight, and passing certain physical fitness criteria. The military was a standards-rich environment. One standard issue I faced as the base commander at Galena was that our one and only civilian barber abruptly quit her job on the base the day before I took command. I had a number of folks ask me if we could just waive the haircut requirement until we got another barber on base. My immediate answer was, "No way... I will not let that standard go." That decision led to numerous barracks barbers trying their hardest to ensure each member of the squadron complied with military grooming standards. There were a few really bad haircuts, but everyone made do and held to the desired grooming standards.

Annual Physical Fitness Exam

The Air Force in 1989 had an annual requirement that every military member was to complete a 1.5-mile run to determine their overall fitness. I did have a few of my older sergeants ask if I could just waive the annual aerobic test since we were all serving on a remote unaccompanied tour. Again, my answer was an unequivocal no, everyone was required to run the annual aerobics test. As an added incentive I gave the entire squadron, all 320 military members, a challenge: anyone who could run the 1.5-mile test faster than the commander — that being me — would earn a three-day pass. Because of the varied work schedules, our administrative support folks set up three separate times to run the test on 16 July. They had laid out the 1.5-mile course on top of the dike that circled the base. I thought I may have over-promised on the three-day pass and would have multiple members taking me up on my promise. The first run was at 0700 hours and there were probably one hundred squadron members at the run. Everyone was skeptical of the forty-two-year-old commander who challenged them all.

Fortunately, I had been running a good bit to deal with stresses associated with my job as a commander, and I knew I could run a fairly good race. In the 0700 running group I managed to beat everyone and finished with a respectable eight minutes and twenty-two seconds. I did wonder if the rest of the runners in this race just let me beat them, but I was one third of the way through my wager. The next run was scheduled for 1130 hours. About one hundred runners were there for the second race. I had passed on lunch that day and still had a good time of 8:27 on this race; again, no one finished in front of me. The final run of the day was scheduled for 1600 hours. I had eaten a protein bar at about 1430 hours and felt pretty cocky about my wager with the squadron. There were about eighty squadron members at this third run. I was more than ready to shine may ass once again. The run was going just fine, I could see the finish line up ahead, and thought I had it in the bag when of my base firefighters passed me like I was standing still. I was very surprised because this individual was a heavy smoker; I had even assigned him to one of the butt patrol teams because a few days earlier he had thrown a cigarette butt in the snow. My time on the third run was 8:20, but I was beat by a good ten seconds and had no choice but to award the fire dog a three-day pass. That was the last time I challenged my young airmen to any athletic contest. But at least the standard of completing the annul Air Force aerobic test was completed.

Alaska Standards

Many times an Air Force base commander can designate civilian establishments in the local communities as "off limits" to military personnel. The reason for the off-limits designation can run the entire spectrum of reasons. While I was in Alaska, the one and only civilian bar in Galena, called Hobo's, was designated as off limits by the previous commander. It seems that a few months before I took command of Galena one of our young airman got in a fight at Hobo's and shot one of the locals. All of the people involved in the shooting were very drunk. The young airman was arrested and charged with felony attempted murder. He had been moved to Fairbanks to await a civilian trial. Even though this event took place before my command, I was concerned about retaliation against my airmen if they ever went to Hobo's. I made a command decision to uphold the Air Force standard, extending the off-limits designation for Hobo's during my command at Galena. Every new person who arrived at Galena during my command was briefed that Hobo's was off limits to all military personnel.

Late one evening I got a call in my quarters at 2330 hours from a very drunk person who said that he was the owner of Hobo's and wanted to talk to the Galena commander. I told him I was the commander and asked if I could help him. He explained that my decision to make

Hobo's off limits was killing his business, and he may have to close his bar. I told him I understood his concern but that I also had a concern for the safety of my airmen and felt that there were still some local folks who wanted to even the score with our military personnel for the previous shooting that occurred at Hobo's. I told him my decision was final but that I might review the decision after the trial of the accused airman was completed. I then received the most eloquent string of profanities I had ever heard as he really exploded about my decision.

This wasn't the last of the off-limits issues with Hobo's. Later in the same month I got some "intel" from a senior sergeant in the barracks that a few of my airmen were, in fact, still going to Hobo's on Saturday nights. I called my security police (SP) chief to my office and told him I needed him to show up at Hobo's on the next Saturday night to see if the rumor about our airmen attending Hobo's was true. He said he would do that task himself and report back to me on the next Sunday morning. Early the next Sunday morning the senior SP reported back to me that he had seen two airmen, other security policemen, at Hobo's enjoying the drinks and the local women. I told him to have both of them report to me early Monday morning. On Monday both sergeants were waiting in my office, and they came up with an incredible story that, on their own, they had heard from the rumor mill that Galena folks were still going to Hobo's and they wanted to catch them in the act. I called BS on both of them and told them that I would begin processing Article 15 actions, a form of non-judicial punishment, against both of them for disobeying my orders. I had already coordinated with Judge Advocate General's Office at Elmendorf and got the initial approval to move forward with the Article 15. Standards must be enforced so that there is no misunderstanding about what is or is not acceptable.

During the final month of my command I was notified that the owner of Hobo's had died from a heart attack, and the locals were so upset that they laid his body right on top of the bar for three days while they came up with a plan for a memorial service and burial. "Only in Alaska," I said to myself.

Civilian Standards

This area of standards was a unique process. Each civilian company that I consulted for had different standards for many activities. From the requirement for written documentation and the format for visual presentation to coordination with competitors and travel requirements, I had to be on top of each area and act accordingly depending on the company I was consulting for.

Complying with these standards led to a crisis during an important presentation to an Air Force two-star general. My client was trying to sell his company's ability to provide support to a very complicated satellite program. This client always told his story very forcibly and really didn't pay attention to the standards of who he was making a presentation to. As I mentioned earlier in this book, I always tried to find out ahead of time the likes and dislikes of the person we were making the presentation to. In this case, the general was a very assertive person and had a standard format that he wanted material presented to him. I explicitly instructed my client about the briefing format the general would expect and told him we may get thrown out of the briefing if we deviated from that standard. True to my expectations, my client forcefully tried to present his material the way he wanted to… not the way the general wanted. I was more than embarrassed. We were unceremoniously thrown out of the briefing and never did get the contract we were hoping for. As I dropped my client off at the airport for his return flight to Boston, I told him that if he didn't start listening to me and following my advice about the standards our potential customers expected he could find another consultant in Colorado Springs. I was as serious as a heart attack on this important issue, and he knew it. He heeded my advice, and I stayed on as a consultant with this company for another ten years; I never had another issue with how he treated our customers. I learned a valuable lesson: be aware of other people's standards and follow them whenever you can.

Personal Standards

I have a good example of when a leader must impose a personal standard on a situation. In 2008 I was the co-chairman of an effort to honor the nearly 300 forward air controllers (FAC) killed in action (KIA) during the war in Southeast Asia. As part of the FAC reunion being held in Colorado Springs, I was tasked along with another FAC to lead the planning, design, and construction of a large granite memorial listing the names of those 300 FACs. This was a huge program whereby we collected nearly $75,000 in donations to build the monument in one of the main city parks in Colorado Springs. We were expecting more than1,500 people to attend the dedication ceremony. As we were rounding up monetary support for the memorial and the memorial dedication ceremony we met with the base commander at Peterson AFB, then a colonel, to request a number of support items from the base, including folding chairs, bus transportation, an honor guard, and many other items that would enhance our dedication ceremony.

His response was that before he would commit any support to us from the base he wanted to know who the guest speaker would be. We explained that retired U.S. Air Force Colonel George Everette "Bud" Day, a FAC himself, prisoner of war (sixty-seven months in North Vietnam), and recipient of the Air Force Cross and Congressional Medal of Honor, would be our guest speaker. The base commander said that he would first have to review and approve Colonel Day's speech before he could commit to support from the base. I nearly went apoplectic in my refusal to submit Colonel Day's comments for review and approval. I "politely" explained that we would find support for the memorial dedication from somewhere else. My co-chairman and I left that meeting not very pleased at all with his response. There was no way I would even consider reviewing what a national war hero was going to say at the dedication ceremony. We eventually found and paid for most of the support we needed for the ceremony from other sources, and I was vividly reminded that sometimes personal standards must be enforced even it costs you in the short term. A good leader must do the right thing every time!

Volunteer Standards

For more than two and a half decades I have led a lay ministry program that pairs trained volunteers from both the congregation and local community to serve others going through difficult times in their lives. We conduct a very intense fifty-hour training program to prepare the lay ministers to provide the best possible support to their care receivers. We established well-defined standards about attendance at our training sessions. Volunteers in training can only miss two sessions during the fifty hours of training. Over the course of twenty-five years of conducting this training, and after training more than 175 lay ministers, we did have to dismiss several potential ministers because they just missed too much class time. Our philosophy was that each person we trained must meet a certain level or standard of training and have the necessary basic tools to care for someone going through a difficult time in life. We did get some strong pushback from those that we would not allow to finish the training, but we thought that the standard of completing the majority of the training was critical to our program.

Supervision Standard

A large part of our lay ministry program involved twice monthly supervision sessions when our lay ministers could get advice and guidance on dealing with their care receivers. Fortunately, we had a very demanding and detail-oriented person, a retired banker, in charge of the supervision sessions. Our program outlined specific procedures to facilitate these meetings so the support to our ministers would be meaningful and we would not break confidentiality concerning the care receivers. Our supervision coordinator followed the process exactly by the book during each session. She would always provide feedback on how we needed to stay on track to effectively complete the process. There was no margin of error and because she insisted on staying within the established standards of our program, we are still going strong at our church with this lay ministry program after nearly twenty-five years. I learned a lot from this banking professional, who was also an expert in processes and standards, about the importance of following the established standards to ensure the success of an organization.

Standards in Leadership Lessons
- Establish standards early in your leadership role.
- Define the exact standards expected.
- Clearly communicate the expected standards.
- Enforce standards at every opportunity.
- Enforce standards equally… no exceptions.
- The leader must meet all expected standards.
- Follow all standards when no one is looking.
- Standards are important in any environment.
- Follow your personal standards every time.

6

INTEGRITY

It is hard to separate integrity from the other foundational tenets of effective leadership. Integrity in leadership is essentially a concept of consistent actions, consistent values, and consistent enforcement of those actions and values. I wrote about the consistency of establishing and enforcing the standards in Chapter Five. A good leader must have integrity while carrying out those standards. When a leader acts with integrity the leader builds trust with the members of the unit, and the entire team will then have confidence in the decisions a leader makes. And believe me the entire team is watching... they know for a fact if the leader is acting with integrity. I have many examples of the integrity factor throughout my entire military and civilian careers. While some of these examples may seem minor, each one builds on the premise that a leader must act with integrity in every action he or she takes, no matter how small.

Combat Integrity

One of the main lessons I learned from flying combat in Southeast Asia was to always tell the truth no matter the consequences. As a forward air controller (FAC) I had the ultimate responsibility to report any bomb damage assessment after an airstrike by the fighter aircraft

I was controlling. At the time in the early seventies the Department of Defense was judging each bombing mission we flew by the number of enemy combatants killed. They were essentially keeping a body count score to justify the continuance of a very difficult and unpopular war. The FAC had the first eyes on the damage following any airstrike. Because I value integrity, I reported as precisely as I could the real number of casualties or equipment destroyed on each bombing mission I controlled. Many times, the only real battle damage assessment (BDA) I reported was the number of trees hit and monkeys killed during an airstrike.

I do know at times that some of the fighters really missed the target that I had designated but still wanted to get a positive BDA for their own score card. I was steadfast in only reporting what I observed. On one particular mission I was controlling a four-ship flight of F-4 Phantom fighters supporting a group of U.S. Army troops in direct contact with the enemy along the South Vietnamese and Cambodian border. The flight had a call sign of Sidewinder 01 which indicated that the lead pilot was most likely the commander of that fighter wing. Sidewinder 01 dropped his first bomb so far-off target that he endangered the ground troops we were trying to support. In all honesty I am not sure his bombs hit the correct country. I immediately called a halt to the bombing mission, sent the four fighters off target and told them that I could not use their support today as we had friendlies in close contact; I would not jeopardize their safety with misplaced bombs.

Just as I expected, when I landed, I had a call waiting for me from the wing commander who I sent off target. He was none too happy to be sent off target for a bad bombing run. I was a fearless first lieutenant at the time and basically told him tough shit because I could not risk having him bomb the friendlies. After a very long silence he said, "OK, you were in charge of the mission, and I probably would have sent me off target as well." I just could not let the fighters continue to drop bombs that were so far-off target.

Flying Integrity

As I progressed in my flying career, I experienced a situation that concerned me a great deal. One of the annual requirements to remain a qualified Air Force pilot was to successfully complete an annual instrument flight check and an annual written flight instrument exam with one hundred questions. As I was preparing for the written exam an old timer pilot in my squadron at Eglin AFB told me, "Don't sweat the exam. Here is a copy of the exam with the correct answers noted... just take the answer sheet with you to the exam." My reply to him was, "WTFO?" (What the Fudge Over) Here was a required test that every pilot must pass every year, and I just couldn't cheat on this test. Instead, I got a non-marked Instrument Written Test and studied all aspects of the exam. I refused to take the answer sheet (we called it the pony) into the exam. While I didn't ace the Written Instrument Exam I did well enough, 85 percent, to pass my first annual test. From that time on I made a point to intensively study for the annual written exam and passed it (without the pony) every year during my flying career.

Beer Integrity

I encountered a unique integrity challenge when I was a young captain flying the C-131 Samaritan out of Eglin AFB. On one occasion I flew thirty-five of the Armament Development and Test Center's head scientists and engineers to Kirtland AFB in Albuquerque, New Mexico for an important conference on the new Airborne Laser Laboratory, a highly modified Air Force C-135 equipped with a laser that could shoot down any aerial vehicle or missile within range of the laser. Later in my career at Hollman AFB in New Mexico I flew multiple photo chase missions with this aircraft.

At that time Coors beer had a cult-like following but the product was not sold east of the Mississippi River. On other flights to the west, I had carried a few twelve-packs of Coors beer back to Florida. On the morning I was ready to fly the thirty-five scientists back to Eglin AFB, my flight engineer informed me that we had a problem with the flight. Apparently, the chief scientist on the trip had coordinated with a local beer distributer to deliver three pallets of Coors beer to the operations flight line with the intent of me flying the beer back to Florida. However, with that much beer and thirty-five passengers we would be way over the maximum weight limit for the aircraft to safely fly. I could just envision the accident report... *Air Force aircraft crashes because it was overweight with a load of Coors beer.*

As the aircraft commander of the flight, I immediately advised the chief scientist that we could not take the beer and the passengers back to Florida because of the weight issue. He told me no problem, he would kick the other thirty-four the passengers off the plane to take a commercial flight, and he would escort the beer himself to ensure it arrived safely. Even as a young captain at the time, I knew this was not an appropriate thing to do as it would cost the government quite a bit of money to get the thirty-four passengers back to Florida at the expense of a few pallets of Coors beer. As politely as I could, I told him that I would not fly the beer back to Florida at the expense of the thirty-four commercial tickets to the Air Force. We had a bit of a stand-off, but he finally realized that flying the beer rather than the passengers was not really a good idea. I recommended to him that maybe the Kirtland AFB Officer's Club would buy the Coors beer from him to avoid the expense of not getting the beer he purchased back to Florida. This was another real-world example of enforcing an integrity issue that a leader knows is correct even if there is some pushback on the decision.

Nike Pencil

One of the annual requirements for all Air Force personnel was to pass the annual aerobics run. This easy test, that I described in the standards chapter, was only a mile and a half. The annual run had to be completed in a designated time based on the airman's age; older service members had more time to complete the run with a passing score. I had been doing a daily three- to five-mile run every day since my return from Southeast Asia so the annual test was certainly no big deal to me. I was amazed when I learned that one of my squadrons used the "Nike Pencil" method to "pass" the annual aerobics test; in other words, they simply wrote in a passing run time. This was not acceptable to me. Whenever I was in a leadership position, I demanded and led the annual run with the people I was responsible for. Remember the story at Galena about this Air Force fitness test and how I challenged my entire squadron to beat the old forty-two-year-old lieutenant colonel in the run.

I used this technique in every situation I led, from the maintenance group in the F-106 squadron to the students I taught at the Air Force Command and Staff College to the T-33 squadron I commanded. I would never expect my people to do something that I couldn't or wouldn't do. The followers are always watching the leader, and once the leader does something unethical it is very hard for the people to trust and follow that leader.

Fighter Squadron Integrity

As the senior flight commander in the 318th Fighter Interceptor Squadron (FIS) I had many responsibilities leading the six pilots in my flight. One of the most challenging responsibilities was to ensure that each pilot in my flight completed every one of their annual flying requirements, including night landings, formation takeoffs, instrument approaches, and many other items on the seemingly never-ending list. At times it was difficult to complete each requirement. One annual requirement was to complete practice intercepts against an aircraft using electronic countermeasure (ECM) jamming.

This was a challenging task because we had to depend on another aircraft, the B-57 bomber, that was assigned to a couple of Air National Guard units for our intercept training flights. During one of our reporting periods the entire B-57 fleet was grounded due to an aircraft accident. We had no opportunity to meet our annual ECM intercept requirements. Toward the end of the reporting period, I submitted all of the requirements that my pilots had completed, obviously without the ECM intercepts. My operations officer asked me to "pencil whip" the ECM requirements as it would not look good to our higher command if we were short of our required ECM intercepts. As a fairly young major at the time, I knew this was wrong and refused to pencil whip my reports. Unbeknown to my me, the operations officer changed my reports and sent them to headquarters. Two weeks later headquarters questioned my requirements report, asking how we completed the ECM intercepts without having the B-57 available. That certainly was a good question, and I told the person from headquarters that I did not submit a report stating that my pilots had completed their annual ECM intercept requirements. When I speculated that someone must have "modified" my report — as I had kept a hard copy of my submission — the doodoo hit the fan. The operations officer was fired for submitting a false report. It was obvious that no one in the F-106 community could have met their ECM intercept requirements since the B-57s were grounded during the entire reporting period. This episode just reinforced the integrity requirement to always tell the truth, no matter the consequences.

Drone Integrity

As I mentioned earlier, I was the assistant operations officer for the PQM-102 Drone program at Holloman AFB. We supported many high-priority missile tests with our unmanned drone. One test was the Weapons Systems Evaluation Program (WSEP) that was supporting a new air-to-air missile, the Aim-9P. The Aim-9P Sidewinder was a variant of an air-to-air missile that was supposed to have a much lower smoke trail when fired. We had a run of bad weather at Holloman that had delayed the test firings of the Aim-9P for almost a week. Based on the forecasted weather delay, my boss, an Air Force lieutenant colonel, had to fly back to Tyndall AFB in Florida for an important meeting. That left me and the other operations officer, an Air Force major, to support the potential drone launch the next day, if the weather cooperated. We had six F-4s and crews ready to fly the mission if the weather was okay. The forecast for the following morning looked good, and we notified the WSEP support team that we would be a go for a 0700 drone launch mission.

At about 0200 in the morning on the day of the drone launch, I got a call from the Alamogordo Police Department that my drone coworker, the Air Force major, was in the drunk tank at the police station, and someone needed to come down to the jail and bail him out. We were only two and a half hours before the morning briefing for the very high-profile drone launch. What a pickle. I immediately went to the police station and saw that there would be no way this major would be in any condition to perform his duties supporting the drone launch later that morning. As a middle-grade captain I had to make a very important decision about the scheduled drone launch. I knew we needed both Air Force officers to safely run the mission, but I also knew the importance of flying this test mission after so many days of weather delays. I had to make a decision well above my pay grade.

With all the courage I could muster I contacted the lieutenant colonel in charge of the six aircraft that were scheduled to fly the Aim-9P test mission later that morning and explained the situation, telling him we would be canceling the mission. I knew my decision would make the excrement hit the fan as the WSEP team had nearly thirty people deployed to Holloman for the Aim-9P test firings. These thirty people included the pilots, the test management team, the aircraft maintenance support personnel, and the munitions support personnel. The teams had been delayed four days already due to the unusually bad weather in New Mexico. The WSEP team leader was certainly none too happy with my decision. I then notified my boss, who was still in Florida, about my decision and the reaction I got from the WSEP team lead. I really had very little choice in my decision; but I also knew from first-hand experience that sometimes the integrity of your difficult decisions pays off in the long run. I say this because later in my career, the lieutenant colonel WSEP Team Leader — a colonel at that point — became my boss in Alaska. He reminded me often of the courage it took for me to cancel his test mission many years earlier in New Mexico. By the way, after two more days my boss returned from Florida and we completed the Aim-9P test firings.

Contracting Integrity

Following my Air Force career I became a consultant for many large defense contracting companies. My job as a consultant was to find military contract opportunities for my clients. The competition for the defense department program dollars was very fierce, and it was not unusual for five to seven different companies to be in the running for upcoming Air Force Space Command, NORAD, or Air Force Academy contract. We would try anything we could do to make sure we had an edge when the contract was awarded. My role as an in-house consultant in Colorado Springs was to meet as often as possible with the Air Force, NORAD, or Air Force Academy personnel who would make the decisions about contract awards. Our competitors did the same thing. Whatever it took to get an edge on the contract award process was what my clients expected of me.

Based on my previous time as the deputy chief of staff of NORAD I had almost unlimited access to any of the staffs of both NORAD and Air Force Space Command. I tried my best not to abuse that position and followed all the normal protocols expected of any other contractor. On one such meeting I was talking to an ICBM expert, an Air Force major, who had worked for me in the past. He was the staff officer in charge of recommending what company should get the contract that I was trying to win for my client. While in his office talking about the upcoming contract, he was called away. As I was waiting for him to return, I could not help but notice a document that was sitting in the trash can next to his desk. I could clearly see the letterhead of one of the competitor companies that was also bidding for the same contract my client was bidding on. Out of curiosity I picked up the document and noticed some cost planning figures that were listed on the paper... our main competitor's cost estimate for the contract. I knew I should not have this information.

As soon as the major returned to his desk, I told him what I had found in his trash can and advised him he should be more careful in leaving sensitive information lying around. I had a real ethical dilemma. I decided right then that I would not divulge the information that I found in the trash to my client... I just didn't think that was the ethical thing to do. As my client went forward with the proposal, I kept my thoughts to myself about their proposed cost of the contract. I did ensure all the technical information in my client's proposal was correct as I knew exactly what the government was looking for as far as technical support. Fortunately, three months later the government accepted my client's bid, and we were awarded a significant long-term contract. My wise decision not to divulge the competitor's cost estimates and pointing out the danger of leaving sensitive material lying around really made my work with this Air Force office much easier, and I had almost unlimited access to this office and their planning staff.

Volunteer Integrity

Following my Air Force retirement I was elected as the president of the Colorado Springs Coalition for Adult Literacy. In this unique role I was part of a team that helped individuals in the Colorado Springs survive with limited literacy abilities. What a real shame when we encountered adults who had graduated from the local public school system who could not read at any grade level. I was flabbergasted that this situation still existed in the U.S. in the mid-nineties. The coalition's goal was to link up the individual with a qualified reading tutor and at least get the individual basically functional to survive in the world. At the time apparently the largest school district in Colorado Springs had a policy that if a student was enrolled in high school, showed up for class part of the time but did none of the required class work they would award that student a grade of a D minus and move that individual along all the way to awarding a high school diploma.

As I further reviewed this insane policy, I interviewed a few of the local companies that hired recently graduated students into entry-level positions. Almost to a person, the consensus was that the company would rather have a student with a general education diploma (GED) than a diploma from the local school district. At least the employer knew that the individual had some basic skills required to pass the GED. I served in the president's position for nearly three years and was so disappointed in the apparent lack of ethical standards the school district was applying that I retired from the board. I know it was difficult to give the students who were struggling extra support from the school but to just pass them along was less than ethical and showed me that sometimes doing the most ethical action is difficult but necessary. The school district was failing those students.

Integrity Leadership Lessons

- Always tell the truth.
- Don't ignore the standards… even from your boss.
- Speak up when you recognize an ethical issue.
- Don't tolerate ethical lapses from others in your organization.
- Always do the ethical thing, *especially* when no one is watching.
- Always make the tough ethical decisions… then stand by them.
- Leaders must comply with the same integrity standards.

7

Loyalty

In any effective organization there is always a level of loyalty between the leader and the followers. Sometimes that loyalty is a one-way street when it is demanded by the leader; other times loyalty is earned by the leader through various actions that enhance the trust and loyalty of the followers. As I observe some of our leaders today there seems to be too much emphasis on the leader of an organization and not the people of an organization. It is the proverbial "me" versus "we" concept.

My experience in leadership roles both as a leader and a follower indicates that in the best organizations, loyalty is a two-way street. In order for the followers to show loyalty to the leader, the leader must earn the loyalty of his followers, which includes showing loyalty to them. Obviously in a military organization there is always the temptation to build "loyalty" through position and fear rather than trust. I have been in those organizations... fear just does not work to build a healthy organization. I have seen too many "yes men" in fear-based organizations. The real information the leader needs to make wise decisions is never forwarded for fear of being ridiculed by the leader.

I have a perfect example of this fear-based organization that I encountered late in my Air Force career when I was the deputy chief of staff of a large international military organization. The top leader of this organization, a four-star Air Force general, was a very abusive and fear-based leader. The entire staff was reluctant to bring new ideas to the boss in fear of being ridiculed in public by the leader if he did not like the new idea. What a difficult situation to be in. The staff had some very forward-thinking ideas to make our organization even better but seldom brought those ideas forward. Hence, the leader made some critical mistakes in guiding our organization. As the senior colonel, I was in a very difficult spot trying to coordinate with the staff.

One humorous situation did occur during this four-star general's first conference with all his region commanders and general officer staff members. For this conference our general wanted to put up an overhead slide about his main objectives as the new commander. He wanted to use a hand-held laser pointer to emphasize each item on the slide. I carefully showed him how to use the laser pointer, which was about the size of a small television remote controller. On the remote was a large red button to turn the laser on and off laser. I told him, "Button in and hold for the laser on. Release the button for off." He tried to use the laser pointer and became so frustrated when the laser kept going off when he released the red button. This frustration turned into fury, and he threw the laser pointer at me. My chair in this particular conference room was about ten feet behind the commander, right next to the conference room emergency telephone. When the "flying" remote control hit a very large oil painting on the back wall, it broke into hundreds of small pieces.

The general screamed at me to get another remote controller that *%#*ing worked. My audio-visual technician brought a replacement controller, and I gave it to the commander and reminded him again, "Button in and hold to turn the laser on. Release the button to turn off." The exact same situation happened again. He couldn't keep the laser on as he kept releasing the red button. He once again turned and threw the remote controller at me. This time I was ready and caught the remote with one hand just before it hit the wall. The entire group of general officers looked astounded at the actions of their commander. As the general turned again and continued his presentation without the laser pointer, I held in the red button on the controller and focused the laser right on the side of his head to show the staff that the laser did in fact work. I could see the other generals sitting at the conference table trying their hardest to keep from laughing. During the next break, two of the generals who saw that my handy laser worked stopped by my office, which was directly across the hall from the conference room, and congratulated me on my fine work!

As I said before, loyalty is a two-way street. I tried my very best to be as loyal as I could to my boss and tried to mitigate the fears of the staff who wanted to bring new ideas to him. I even went so far as to try to pre-brief the boss on potential new ideas that the staff had come up with... to no avail. I worked my hardest to assure the staff that the leader was not as bad as he appeared. Again, every time I tried to cover for the boss, he would have another outburst and completely destroy my attempt to paint the boss as a "good guy." I was so happy that I was at the end of my career and only had to put up with the buffoonery of this leader for a little over a year. I certainly learned a leadership lesson in this situation. A leader has to earn the loyalty of his followers. On the positive side of the loyalty equation, I have many stories of just the opposite as the story above.

Another story about this general officer really solidifies the two-way aspect of the loyalty piece of leadership. You may think I am just picking on this general, but he gave me so many insights into how to not act as a leader that I am appreciative of his negative role model status. It seems that the general had made quite a name for himself in his previous assignment and the first three months in my organization, and he was under investigation by the Air Force Office of Special Investigations (OSI). I was called to testify about what I knew regarding the various charges against this general at his previous assignment and his first three months at the new command.

During the three intense hours of interrogation, I tried my very best to stay to just the facts about what I knew about the general's activities since he arrived at the new command. I was very honest and told them exactly what I knew but did not share any of my personal feelings about the general. Once the three-hour interrogation was completed, I was told that a copy of my confidential testimony would be sent directly to me in a sealed envelope once the investigation was completed. After about a three month wait, I was called to the general's office and told I had a package I needed to pick up. This was not an unusual request as I saw every completed staff action package as part of my duties as the deputy chief of staff. The general's executive officer gave me a package that was marked sensitive and had a tape over the opening. The executive officer told me it was a copy of my testimony. I could readily see that the confidentiality seal had been broken, and I can only assume that the general had read my confidential testimony. So much for a two-way loyalty check. I must have passed the general's warped sense of loyalty because I was not fired on the spot.

Flying Loyalty

The best approach to flying aerial combat is to rely on the strength in numbers. As pilots, we would normally train and fight in two to four-ship flights. For every flight one of the pilots would be designated the flight lead. The wingman or wingmen in these flights had complete loyalty to the flight lead as the designated leader to direct all the actions necessary to have a successful aerial engagement. The flight leads that I flew with were always the most experienced or competent pilots in the squadron. They had demonstrated through their flight leadership skills that they could direct a flight of two to four fighters in the best possible scenario to be successful; as wingmen we had complete confidence in their ability to lead us properly and felt a sense of loyalty to follow exactly the directions given by the flight lead.

The loyalty shown to the flight lead by his wingmen was an earned loyalty. A wingman had to maintain situational awareness as to what was going on a three-dimensional setting. The more skilled the flight lead was the more the wingmen trusted their ability to lead the engagement with the enemy or the friendly adversary in a peacetime air combat training (ACT) scenario. This kind of loyalty transcended the flying business. If I had confidence in a leader, I would be comfortable follow him wherever he led us.

Team Loyalty

Part of my leadership loyalty equation is recognizing the good work and support of your followers. While I was the director of readiness for the Alaskan Air Command (AAC) part of my duties was to provide all the simulation activity during our NORAD Alaskan Region (ANR) training and evaluation exercises. I had a six-man team that was responsible for every aspect of the simulation process. They simulated everything from exercise defense conditions (DEFCONs) and simulated Soviet air attacks to simulated biological and nuclear attacks against Alaska. The team was so talented and professional that just before a major audit and evaluation, representatives of the NORAD Inspector General (IG) requested that my simulation shop set up all the simulation activity for this event. I told the team to set up a very challenging simulation that would really test our Alaskan NORAD Region (ANR) battle staff and ANR radar controllers to the fullest. After all, I had directed that all previous simulation practices be at the most challenging level possible. I did get some feedback from the Alaska NORAD Region personnel that my simulation shop was way too tough on them during the training exercises.

The goal of this annual audit was to evaluate every action the NORAD region staff would take under simulated wartime conditions. The region was evaluated on the actions of the battle staff, coordination with the supporting units, the transition from peacetime to wartime, the simulated deployment of forces, the enforcement of the rules of engagement (ROE), information flow, assessment of intelligence, and reconstitution after an attack. This was a huge deal and really tested our forces on how they would react under a wartime scenario. Because my exercise shop had tested the NORAD region staff so intensely in the previous years, we were not surprised that following this audit evaluation our region was awarded the first ever "Outstanding" rating, the highest rating possible. In the final report the NORAD IG team specifically mentioned the skill and hard work that my simulation shop had provided to the evaluation process.

The NORAD region commander, an Air Force three-star general, was ecstatic about the incredible score we had received, and he directed that the entire battle staff and all of the NORAD region weapons controllers be awarded an Air Force Achievement Medal for their outstanding performance during. I fully supported that decision but questioned why my simulation shop was not also awarded the Air Force Achievement Medal. While only a lieutenant colonel at the time, I brought that issue up to my boss, a colonel, who brought it up to the general. The general agreed, and my entire simulation shop was also awarded the much-deserved medal. I knew how much hard work the simulation shop had put into the preparation and training for this critical evaluation as well as the hard work they did to support the IG team during the actual event. I found that whenever you reward your people for outstanding work you are not only recognizing their loyalty to the mission but also showing that loyalty is a two-way street from the leader as well.

Follower Loyalty

As I stated above, loyalty is a two-way street. The leader must always show loyalty to his followers in order to gain the loyalty from his followers. I like to use the story of how I built the loyalty of my followers when I was the official leader of two different organizations. I had worked for too many bosses who would only ask a question that they already knew the answer to. I am sure they were trying to test the knowledge of the person, or they were so insecure that they were afraid to ask an honest or legitimate question thinking the folks would view them as dumb or lacking. I made it a point during my very first in-briefing to the people assigned to my organization to explain that I would ask a million questions because I did not know everything. I could feel the skepticisms of my folks when I told them that. But I really tried to follow that model. I asked a question whenever I really didn't know the answer.

This formula worked when I was the commander of a flying squadron. As I was new to this mission in Alaska, I truly wanted to know why we did some of the things that seemed normal to the old timers. Many times, I would be in the "duh" mode as the answer they gave me made perfect sense as to why the procedure or action was done that way... I truly did not know the answer. Over time, I believe the squadron personnel started to trust me and would not be on the defensive whenever, which was quite often, I asked a probing question. I started to feel the loyalty of the squadron coming together, and I started to make some very informed decisions as to the direction the squadron was going.

Assignment Loyalty

As my time as the commander of the flying squadron in Alaska was drawing to a close, I was notified that the entire squadron's mission was to be retired and the eighteen airplanes would all be sent to the "boneyard" in Tucson, Arizona. The boneyard is the U.S. Air Force aircraft and missile storage and maintenance facility located on Davis-Monthan AFB. The formal name of the boneyard is the 309th Aerospace Maintenance and Regeneration (AMARG) Center and takes care of nearly 4,000 out-of-service military aircraft from all branches of the U.S. government. The arid climate of the region makes the 309th AMARG an ideal location for storing aircraft as there is very little humidity that would corrode metal.

I had a difficult task of making sure all eighteen of the young pilots in my squadron got the best possible follow-on flying assignment. I knew that in order to do that I would have to rate each pilot's flying ability and place them in rank order (1-18) against of all the other pilots in the squadron. I had flown with many of these pilots during the previous year but wanted to get a formal, recognized process, in place for me to honestly rate each pilot. I coordinated with my operations officer and had him schedule me for a back-seat flight with each of the eighteen pilots over a two-month period. I had a predetermined checklist, in my mind, that I would use to evaluate each pilot. The criteria of the evaluation included briefing ability, flying skills, and overall situational awareness and control of the flight. Each one of these evaluated flights involved a dissimilar air combat training (DACT) flight simulating aerial combat against the two F-15 squadrons that were also assigned in Alaska. Not only was the flying challenging but it gave me a great deal of insight to help determine which type of aircraft each one of my young pilots should be assigned to following closure of my squadron. Ranking those eighteen pilots was a difficult task.

Once I had completed all the flights and had my rankings set up, I thought it was only fair that I sit down with have each of the eighteen pilots individually to discuss my honest and straightforward analysis of their performance and why I placed them in a particular order. Remember, I had built their trust over the year through my honest and straightforward questions. I believe they trusted my judgment. While it was easy debriefing the top ten on my list, I did receive some feedback from a few of the folks who were toward the end of the ranking list. In one particular case the pilot told me he just had a bad day on his evaluation flight. I said, "OK, how about another flight to show me what you can really do?" The second flight was much better, and he did move up one slot on my rankings. The good news in this entire process was that the eighteen pilots respected my evaluations and each one of them was assigned to a top-line fighter — the F-15, the F-16, the A-10, or the F-4 — for their next assignments. I was more than pleased with all the follow-on assignments and felt that my straightforward approach had cultivated loyalty in the squadron, and the eighteen pilots were, for the most part, satisfied with their follow-on assignments.

Command Loyalty

I tried to use that same loyalty formula that I used as a flying squadron commander when I became a base commander in Alaska. My main expertise up until this time in my Air Force career was mainly in operational flying, flight testing, and aircraft maintenance. I had a legitimate reason to ask many questions about those areas that I did not have hands-on experience with. So, during every in-briefing of every military person assigned to the base, I stated that I would be asking a million questions about their area of expertise and that there was no hidden agenda in the line of my questioning. I truly wanted to know why we were doing a procedure, a test, or an activity in a particular way. Once again, it took some time for the squadron folks to realize I was just asking questions to get an answer on why or how.

As I dealt with many difficult personnel issues as the base commander, I relied on the fact that I had established exact standards of performance for every member of the squadron. I knew that if I honestly and faithfully enforced those standards the squadron members would respect me and support some of my difficult personnel decisions — from the requirement that all military personnel would have a military haircut, even though for two months we did not have a civilian barber on base, to demanding that everyone successfully complete the annual Air Force required aerobics run. I believe enforcing those standards helped the squadron trust my decisions and showed some level of loyalty to me.

Two concrete examples of the loyalty factor came on two very serious personnel issues. I had to show loyalty to all the members in our squadron by adhering to the standards that I insisted on maintaining. The first involved an alleged rape of one my female food services staff members by a male member of my military police force. I knew beforehand that alcohol was involved in the accusation. But I still had to enforce the standards because I knew that the fifty women in the squadron would be watching my every move. I referred the case to the Judge Advocate General's office in Anchorage and they proceeded with the court-martial trial. After a three-day trial it was determined that the incident in fact was consensual and that alcohol use by both parties was involved. The court-martial board found the defendant not guilty of the rape. After the defendant and the claimant returned to Galena, I sat down with all fifty women in my squadron and explained in detail what took place in the court-martial. The women weren't very happy with outcome but were pleased that I brought charges for the alleged rape. I do believe the women in my squadron respected me for at least going forward with the charges and then letting the legal system make the final decision.

The second personnel issue involved one of my young captains who had been accused of violating some critical promotion test standards. I knew all my enlisted folks were watching what action I was going to take against the officer after I brought rape charges against one of my enlisted members. I consulted again with the Judge Advocate General's office and determined that the best course of action was to proceed with a court-martial for the test violation. Two months later the court-martial convened and found the captain not guilty of the charges. Once again, the squadron determined that at least I proceeded with the trial and did not just put that standard under the rug because it involved an officer. This another example of the importance of loyalty and ethical trust.

Loyalty to Your Boss

This can be a very difficult task for a follower and a leader if you work for a boss who doesn't earn the loyalty of his followers. I have described in some detail earlier how difficult it was for me to be to be loyal to the four-star general who treated his staff so terribly. Despite the poor leadership skills of the general, I tried my hardest to be loyal to him. I learned these lessons along the way and applied them to many of the people, good and bad that I worked for. I will list a few of the techniques I used to support loyalty to my boss. Not all were successful but for the good leaders these actions were appreciated.

No Surprises

I tried to give my bosses a heads-up whenever there was incoming information that would affect the mission we were trying to do. The bottom line was that I did not want my boss surprised by something. My first real experience in this technique was during my flying combat in Southeast Asia in the early seventies. I had a unique situation occur on 1 November 1971 where I needed to keep my boss in the loop on a mission I had completed. I had flown a very intense combat search and rescue (CSAR) mission where our U.S. Army Special Forces had retrieved the body of a pilot who had died in an F-4 crash just inside the border between Vietnam and Cambodia. I had worked all the details of the body recovery and had directed that body be flown back to our Special Forces base at Quan Loi. This very remote base had no refrigeration capability, and I wanted to make sure the body was flown back to the morgue in Saigon. I requested that one of the recovery helicopters fly the body back to Saigon. The pilot of the helicopter told me the weather was too bad to the body back.

Being the impetuous lieutenant that I was, I decided that I would fly the body back to Saigon in my FAC aircraft. I loaded the body on the back seat and proceeded to fly to Saigon in some very terrible weather conditions. For a while I thought I was going to crash trying to fly through the very large thunderstorms that I encountered. The other FAC pilots at Quan Loi were concerned that I flew in such bad weather conditions and contacted my boss who was in Saigon for a meeting. After a heart stopping flight, I made it safely to Saigon and had the U.S. Army morgue pick up the body. As soon as I could I contacted my boss and told him of the harrowing flight and why I had chosen to fly the body back in such terrible weather conditions. I explained that I just couldn't stand the thought of this brave pilot's body remaining overnight in a non-refrigeration condition. We had a duty to honor our dead. My boss recognized what I had done and was okay with my decision but advised me to be more careful in the future. Letting my boss know the reason for taking that dangerous flight solidified my concept of always keeping your boss informed and not giving them any surprises.

Boss Heads Up

When I was the deputy chief of staff of NORAD, I made a point that every single document that went to the four-star general for review and or approval included a concise summary written by me with a recommendation on how he should proceed with the issue. Many times, I would disagree with the generals on the staff who sent issues forward and I felt that I had a duty to let the boss know the potential problems that could arise if he took a particular recommended action. I would not always convince the boss to take my recommendations but felt that I should give him as much information as possible to support a wise decision. Despite the bad leadership skills of this four-star general, I believe that I was doing the correct thing in giving the boss the best information possible to make a decision.

More Boss Heads Up

I had another humorous event happen in the first week that I was a faculty instructor at the Air Command and Staff College. The first Friday of class several old friends from Vietnam got together at the officer's club to celebrate that it was the end of the first week of classes. Since I knew several of the new students, I joined with them in the celebration. After several hours of merriment and heavy drinking it was time to go home. As I started to walk to my on-base quarters, I noticed that one of the students in my seminar was being detained by the security police. He was stopped for urinating in the bed of the security police pickup and was being transported to the military police building to be charged with indecent exposure. I knew that this was something I would have to deal with first thing on Monday morning. Fortunately, for me, I saw the commandant of ACSC, a brigadier general, at the morning chapel services that Sunday. In the most apologetic tone I could use, I informed him of the incident by my student the previous Friday night at the officer's club. While he was less than pleased with the incident, he did thank me for giving him a heads up and told me to bring the student to his office the next day at 0700 hours. The commandant came up with a plan that put my student on notice that one more screw up and he would be removed from the class. I learned a valuable leadership lesson with this incident: always tell your boss what is going on, whether the information is good or bad information, so that he or she is not surprised.

Volunteer Loyalty

I had another interesting case in my volunteer role as a Christian counselor. One of the lay ministers that I had trained had been arrested on a very serious felony charge. This charge certainly could have had serious implications for our church and our senior pastors. Just like the ACSC urination story, I immediately contacted our senior pastor and explained what I knew about the situation and together we came up with a plan to not only deal with and support the individual but also to deal with the potential fallout among the congregation, which there was plenty, especially if the facts of the incident were released to the general public. Once again keeping your boss in the loop is a very wise thing to do.

Civilian Loyalty

In my post-military career as a defense industry consultant, I worked with many different companies, from large Fortune 500 firms to small, two-person startups. I was always very mindful to separate the activity between the different companies that I represented. I really had to balance my activities to remain loyal to the companies that were paying me for my expertise. I had many opportunities arise where I certainly could have been effective advising company A on a contract proposal but had already committed my expertise to company B for the same contract. I had some interesting discussions with the multiple companies about who I was exactly consulting for. I never once broke the loyalty to the primary company that I was supporting even with the enticement of a huge payday for my expertise.

There were a few cases where my expertise was on the fringes of a couple contracts that multiple companies were chasing. In those cases, I discussed my proposed actions with the various companies and got their permission to consult for two or more companies. One interesting contract situation involved the development of all future NORAD plans for the next twenty years. Since I had been the deputy director of NORAD for planning, I was a highly sought-after source from the company that was awarded the contract to develop those future plans. I coordinated with my current three client companies to make sure they would have no problem with me consulting for a fourth company. I felt that I would be a very valuable asset to all of the companies I consulted for regarding future work if I knew what NORAD's twenty-year plans would be.

In a few cases I saw how advantageous it would be if a couple of my client companies collaborated to formulate a much better contract proposal for the government. I thought that approach would maintain my loyalty to each client company while enhancing the probability of both companies winning part of the contract. This approach was not always successful even after I got the two companies to team together on a contract. In one case I found out through my military source that it really didn't matter who submitted a proposal, the current contract holder would be awarded the new contract. In this situation I assertively advised the two companies that teamed together that they should not even bid on the contract as the rumors were very solid that another company would be awarded the contract, no matter what. It took a couple of these "inside" information contract scenarios to convince my companies that I was giving them the best information available to make a wise bid proposal decision. I felt that each company trusted my expertise as I showed a great deal of loyalty to them.

Volunteer Loyalty, Part 2

As the president of the Colorado Springs Coalition for Adult Literacy I had an opportunity to coordinate with many of the city's charitable organizations and had a good rapport with all of them. As we progressed with our mission of helping adults become literate enough to function in society, the State of Colorado established a similar adult literacy program for additional help to teach these at-risk adults to read and write. The year before this new program was established, I had coordinated with the local United Way Foundation to add the coalition on the donation list. We were very successful, and many people in the community donated money to our literacy program throughout the year. As the state literacy program started to take hold our small local program became obsolete. My staff was beginning to get frustrated with the lack of action that our organization was doing to help with adult literacy.

As a result, I called an emergency meeting of the Coalition for Adult Literacy Board to request their opinion on what we should do next. There was a unanimous consensus that we should retire our local literacy program and transfer all of our activity and funds to the state literacy program. I learned that if a leader does the right thing, in coordination with the staff, that leader can maintain the loyalty of the organization. In this case the board was more than pleased that I suggested closing our literacy activity. We were also able to easily transfer our United Way funds to the state of Colorado literacy program.

Loyalty Leadership Lessons

- You will always have a boss.
- Don't let your boss be surprised.
- Loyalty is a two-way street.
- A good leader earns the loyalty of his followers.
- Loyalty sometimes requires difficult decisions.
- Loyalty requires maintaining standards in every situation.
- Recognize and reward the work of your followers.

8

DECISION MAKING

Leadership and Decision Making

In the previous chapters of this book, I covered many of the foundational aspects of my leadership model, from competence and mission to people, communication, and standards, to ethics and loyalty. Not one of these leadership aspects can be implemented without, in many cases, decisive decision making by the leader. In this chapter I will relate different scenarios where decision making was the key factor in implementing or causing an action to take place. These short stories will cover my experience in twenty-six years of flying eight different Air Force aircraft and flying 535 combat missions in Southeast Asia, where my decisions meant the life and death of others, to some of the more mundane leadership decisions I made in command and staff positions, and also my experience in the civilian world as a defense contractor and volunteer leader.

In each one of these scenarios, I had to 1) identify the most critical factors that could affect the overall outcome of my decision, 2) make a timely decision, sometimes within seconds, and 3) be prepared to live with the consequences and potential uncertainty that my decision might lead to. Not all of my decisions were perfect as I don't believe there is ever a perfect decision in any action a leader may take. My experience is that you have to make your decisions based on the very best information you have at the time. Delaying a decision is no decision at all and can lead to some very unintended consequences. I also have the philosophy that if I get better information following my initial decision, I have no problem changing my decision based on the new information. A decision delayed is a decision not made.

Flying Decision Making

It is quite intuitive that flying an airplane involves many critical time-sensitive decisions on the part of the pilot. I learned this lesson from the very first flight I took in the Air Force at Reese Air Force Base in Lubbock, Texas in the late sixties. While initially learning to fly an airplane challenged my decision-making abilities, I had to process and make thousands of critical decisions during each and every flight. The quality of those decisions determined whether I would safely return from the flight. As I became more skilled in flying, some decisions became automatic. The decision process, however, was still in place throughout each flight I made.

While flying the airplane became almost second nature, piloting a supersonic jet fighter presents an almost continuous set of decisions that must be made to safely conduct the mission. In addition to the multiple decisions of just flying the airplane, I had to make decisions on the other members in my flight, the actual mission itself, the potential threat environment, the assigned airspace, the weather, potential emergency situations, the amount of fuel remaining, and a thousand more factors that had to be continuously considered.

The Air Force had a standard procedure for any pilot to deal with and make the appropriate decisions whenever there was an in-flight emergency. Even thirty-two years after my last Air Force flight, I remember exactly the steps to take in the decision-making process during an in-flight emergency:

- Maintain aircraft control.
- Analyze the situation.
- Take corrective action.

I believe this flying decision-making process was a real foundational skill I learned for the many more decisions I would have to make in my Air Force and civilian careers. Flying showed me that decisions sometimes have to made immediately and can have real world consequences if the decision is wrong. This skill of decision making was never more important to me than during my one year of flying combat missions in Southeast Asia during the Vietnam War.

Combat Decision Making

As I explained earlier in this book, one of the most challenging missions I was assigned in the Air Force was as a forward air controller (FAC) in Southeast Asia during all of 1971. I won't rehash the mission of the FAC here, but the decision-making process was very intense on every single one of my combat sorties. I had to decide if I was using the correct map, if I was talking to the correct ground troops I was providing air support to, if the fighter aircraft saw the target that I marked, if the correct munition was being used … there were multiple and sometimes overlapping critical decisions that I had to make. These decisions had to be made instantaneously in some cases to ensure the safe outcome of the mission. To describe that decision processes, I'll recount two specific stories that I described in my first book, *Pretzel 06: Memories of a FAC in Southeast Asia 1970-1971*.

FAC Mission in SEA

Following my in-country combat-ready check and short time as a FAC for the U.S. Army 25th Infantry Division, I was informed that I would be flying in support of a program called the Military Assistance Command (MACV) Studies and Observation Group (SOG) or MACVSOG. U.S. Military Assistance Command, Vietnam (MACV) was a joint-service command of the U.S. Department of Defense. MACV was created on 8 February 1962 in response to the increase in United States military assistance to South Vietnam. The mission of the SOG group, and my FAC unit in particular, was to support the insert, control, and extract of small Special Force teams into Cambodia and Laos. These teams were comprised of U.S. and Vietnamese Special Forces as well indigenous Montagnard troops and Cambodia Special Forces personnel. In 1971 the U.S. was attempting to teach the South Vietnamese (SVN) Army Special Forces to do the SOG mission on their own, and some of the missions were conducted entirely by the SVN SF and indigenous forces. On these particular SVN SF missions we carried a Vietnamese translator who could communicate with the teams on the ground. It was always a challenge taking someone in your cockpit who barley spoke English and who usually got airsick on every flight.

Some of the missions included covert area reconnaissance, traffic watching, capture of enemy forces, destruction of enemy supplies, and, in some cases, the assassination of key NVA or VC leaders. SOG's activities were intended to counter North Vietnam's use of Laos and Cambodia as vast staging areas and supply routes that were supposedly immune from attack. Since the U.S. government denied we had any troops in Cambodia and Laos, even as a lieutenant I could see why this mission was so classified. Following the initial briefing from my air liaison officer (ALO) and the Special Forces troops, I was required to sign a non-disclosure agreement that I would not tell anyone about the mission we were doing for the next twenty-five years. Wow, I really was in a spooky business. You can imagine the various decisions that I had to make on each and every flight supporting this top-secret mission.

One of the unique missions to support the SOG was a crazy High Low mission that we flew along with the 5th Special Force Aviation Battalion and other SOG personnel. Since there were few actual aerial reconnaissance assets available to fly missions in Cambodia, the SOG team came up with a way that we could do our own organic photo reconnaissance flights. These flights involved the use of an Army 0-1, a very small single engine plane, flying at tree-top level with an enlisted SOG combat photographer in the back seat taking pictures of anything that looked like a war target. I would fly my FAC aircraft about 500 to 1,000 feet above and behind the 0-1 with a 1:50,000 map spread out in my lap numbering each mark that was called out by the photographer. Following the flight, the photos were developed and matched up with the corresponding numbers on my map. This way we could match an exact location of any interesting equipment or activity to a precise geographical location.

Unbeknownst to my FAC squadron headquarters at Phan Rang in South Vietnam, we carried defensive rockets on these High Low missions. The rockets were either seventeen-pound high explosive or flechette rockets and were to be used in an emergency if the low bird on a High Low mission was taking fire and needed to exit the area in a hurry. I preferred to carry one pod of the HE and one pod of the flechette rockets. That gave me seven rockets in each pod on opposite wings. During my ten months as a SOG FAC, I completed forty-two High Low missions and every one of them brought a great deal of excitement.

One specific High Low mission on 14 July really tested my skill and cunning, and especially my decision-making ability. On that day we were tasked to take photos along the Mekong River near the town of Kratie. We had known for quite some time that Kratie was being used by the North Vietnamese Army (NVA) and the Viet Cong (VC) as a storage center for a lot of their supplies. We later found out that the supplies stored in and around Kratie was for the NVA invasion of South Vietnam in 1972. We were determined to find out how they were getting their supplies into Kratie. I took off at 0400 and flew down to Saigon to pick up the enlisted photographer, a Marine gunnery sergeant, and returned to Quan Loi just at dawn to brief with the 0-1 pilot. Our plan was to be on the river in the early morning hours and take photos of any activity that could indicate how those supplies were getting into Kratie. Fortunately, the weather was clear with no low-level clouds so we would have a good opportunity to get quality photos if anything was taking place. The 0-1 flew very low, almost at water level, right along the bank of the river.

We planned to fly from south to north and pass the main part of Kratie at about thirty minutes into the low-level flight. I was positioned above and behind the 0-1 at about 1,000 feet. I had to do a bit of maneuvering to keep the sun to my side because it was rapidly rising in the east. I also needed to keep the 0-1 in sight at all times and have my maps readily available to annotate each mark that was called out. I had done enough of these High Low missions that I knew I needed to have my maps prepositioned and sequenced for easy access as we flew the reconnaissance route along the river. We were about fifteen minutes into the low-level portion when all hell broke loose. Just as we passed a small tributary river that flowed into the Mekong the low bird called out that he was taking small arms fire just north of the tributary. I had already armed both wing rocket pods before we descended for the mission and immediately fired off two flechette rockets into an area I guessed the ground fire was coming from. I could not see the actual shooters, but my goal was to at least get the bad guys to stop shooting so the 0-1 could immediately escape. I saw the 0-1 make a hard left turn over the river, and I believe I saw water splashing as he tried to get away from the ground fire.

At this point I could see about ten soldiers shooting at the 0-1, and I then shot four of my high explosive rockets from the left wing pod. When a 2.75-inch folding fin rocket is shot it gets to supersonic speed before it hits the ground. The sound of the incoming rocket is usually enough to get the bad guys to keep their heads down. In this case the bad guys just kept shooting at the 0-1 and then at my aircraft. In a last desperate attempt to get us out of there safely I salvoed all my remaining rockets into where the fire was coming from. I watched as the 0-1 completed his turn over the river and flew southeast back toward the opposite riverbank and away from the ground fire. I made a very hard climbing right turn and headed away from the riverbank area.

Finally, the shooting stopped, and I spotted the 0-1 climbing back to altitude. We both climbed to about 1,500 feet, and I joined in close formation with the 0-1, checked his aircraft for any battle damage and asked him to look over my aircraft. It was obvious from my observation that the 0-1 had taken a few hits. Most of the damage I saw was on the rear part of the fuselage behind the cockpit area. The 0-1 pilot said his engine was running fine and that he was headed back to Quan Loi. He said he did not see any obvious damage to my aircraft, and I followed him back to Quan Loi for a safe landing. Once on the ground we both inspected our aircraft again; the 0-1 had seventeen holes from small arms fire in it. My 0-2 aircraft had three holes from small arms fire in the right tail boom. Taking hits in our aircraft earned us the nicknames of magnet asses!

Typical Air Strike Mission

The best way to describe these events is to relate one air strike mission that I supported on 30 July 1971. My decision-making process was on steroids during this mission, an immediate airstrike following contact with a fairly large enemy force by one of our inserted teams. The team had been sent to the field to observe a potential storage site of enemy supplies and were in the field only two days when they encountered a contingent of NVA/VC forces. The insert had gone smoothly, and all the radio calls were uneventful until the team gave an emergency update at the 1800 hours check in on the 30 July. The team radioed that they had spotted the storage site but had been detected by the enemy forces nearby. They were heading to the emergency landing zone that had been identified prior to the start of the mission. I immediately called a Prairie Fire Emergency, which signified that a team was in trouble and needed immediate air support and extract. I contacted the airborne command and control C-130 aircraft — the call sign was Ramrod — and asked for any air support available to assist in the Prairie Fire Emergency. Simultaneously I called back to the operations team and requested that the extract team along with four Cobra Gunships be put on strip alert, meaning they would be ready to launch on my request.

As it was getting toward the end of the day and sunset, I wanted to get the fighters on target as soon as possible. Ramrod informed me they had diverted two A-37 Dragonfly attack aircraft, Rap 21 and Rap 22, that would rendezvous with me in about twelve minutes. The first thing I had to do was sort out where our team was located, the direction of travel they were taking to get to the landing zone (LZ), and where the enemy forces were located. This contact was done on a dedicated frequency modulated (FM) radio. I could tell the enemy was fairly close as I could hear the anxiety in the team leader's voice. Since I could see that the team was less than one kilometer from the LZ, I advised Quan Loi to launch the strip alert birds, Green Hornet 33 and 34, and the four Cobra Gunships, call signs Raider 21, 22, 23 and 24. They would have about a thirty-minute flight to the LZ.

I had finally spotted the enemy troops about 300 meters behind the team, and it looked like they were closing fast. It seemed to me like there were about thirty of them. Just then Rap 21 checked in with me and advised that they had me in sight and would orbit at 10,000 feet. They had about thirty minutes of fuel remaining, and each had three 500-pound MK-82 bombs and three CBU-24 cluster bombs in addition to several hundred rounds of 7.62 ammo mounted in a nose gun. Before any air strike took place, the FAC briefed the fighters on the parameters of the mission. Some of the material briefed by the FAC to the fighters included:

- Target description
- Target location

- Desired results
- Target elevation
- Location of friendlies
- Landmarks
- Run in heading
- Break heading
- Bailout direction
- Enemy location
- Ground fire expected
- Location of guns
- FAC location during strike
- Desired sequence of bombs
- No bomb line

Once all the briefing material was covered and the fighters acknowledged they understood, it was time for the show to begin. I would call "FAC in for the mark," meaning that I would attempt to place a white phosphorous (WP) rocket exactly where I wanted the first bomb to hit. If the rocket and I performed as desired I would clear the first fighter in with the call, "Hit my smoke," meaning to drop the first bomb where I had dropped the WP rocket. Most of the time the first bomb was dropped fifty meters or so to the north of my smoke. I would then call, "Cleared hot," which meant the first fighter could proceed with the attack and that I had cleared the fighter to expend their ordnance. On this first pass I requested the 500-pound MK-82 bombs on target. I felt that the large explosion from those bombs would force the enemy to keep their heads down. As soon as I saw the first bomb hit the ground, I adjusted the second fighter to hit either on the same spot or about fifty meters to the north. During this mission both fighters hit exactly where I needed them to hit. While there was a large number of small arms rounds fired at the fighters and at my aircraft, none of the rounds hit any aircraft. I checked with my team on the ground to make sure the bombs were not too close to them. In this case my ground team said the shooting from the bad guys had subsided a bit and they were nearly at the backup LZ. Just as that call was confirmed the extract team made radio contact with me and said they were five minutes out.

I called the fighters and told them to make the second pass in the same direction and drop on the same location, announcing another "cleared hot" call. After I saw the second bomb from the flight lead, I told number two to drop in the same location and then hold for the insert of the extract helicopters. I wanted the fighters to drop their last 500-pound bombs about one hundred meters to the southeast of their first two bombs and then follow up with CBU while the extract was taking place. There were no small arms fired during the second A-37 pass. Fortunately, there was a very prominent dry creek bed between their first drop location and the LZ, and I directed no bombs southeast of the creek bed. We called this a bomb line, and that gave the fighters a solid visual clue to not drop any munitions on a designated side of the line. In this case the dry creek bed was the bomb line. I then got a call from the extract helicopters and the Cobra gunships that they had me and the A-37s in sight. I immediately contacted the team on the ground and told them to proceed to the backup LZ for pick up in three minutes.

I briefed the extract helicopters and the gunships about my plan using the very high frequency (VHF) radio channel. I wanted the backup extract helicopter to orbit at 1,500 feet southeast of the LZ while the primary extract helicopter landed to pick up the team. I directed the Cobras to fly escort on either side the extract helicopter and strafe both sides of the LZ as the extract chopper landed in the LZ. Simultaneously, on the ultra-high frequency (UHF) radio channel, I cleared the A-37s to drop their final 500-pound bombs on their previous target and then follow up with two CBU's in the same general area. Since the A-37s' target was a good 500 meters northwest of the LZ, I cleared them in hot for their bomb and CBU drops. That part of the plan worked perfectly, and I am sure the enemy was hunkered down and or killed in the subsequent drops. I then descended to tree top level to mark the LZ with the standard "Bingo, Bingo, Bingo" call. The lead helicopter acknowledged that he had the LZ in site as was prepared to land and pick up the team.

The extraction went as planned with no ground fire noted. Before the A-37s and the extract helicopters departed the area, I requested the four Cobras remain on station while I did a bomb damage assessment (BDA) on the bomb and CBU drops. A BDA was a procedure whereby the FAC would brief the fighters on the effectiveness of their munition drops. We would tell them how long they were on target, the percentage of munitions on target, and the damage that they caused. We always added a body count of enemy killed in action (KIA) or wounded in action (WIA). On this particular BDA I spotted about sixteen NVA/VC bodies near the bomb craters from the 500-pound bombs with two more bodies near the CBU drop. I reported the BDA on to the A-37s as, "All bombs on target, eighteen enemy troops KIA, and unknown number of WIA." I also relayed that the team was successfully extracted and thanked them for a job well done. This was just a standard day in the life of a FAC and a good summary of the many difficult and timely decisions that had to made to ensure a successful mission.

Career Decision Making

You would be correct to assume that most career decisions in the military are made for the service member, with little say or input from the individual. For the most part that was true for my military career; however, there were some key career decisions that I made along the way that really helped my career as a leader.

While I was assigned in Vietnam my primary career question was about my flying assignment after I left the war zone. I started to ask around to see where the other lieutenants who were finishing up their one-year tour in the war were being assigned. I certainly did not like the answer I got: most of the lieutenants were being assigned to B-52 bomber squadrons or the KC-135 tanker squadrons. I did not want to fly a multi-engine or multi-crew aircraft. So, impulsively, I submitted my paperwork to establish a date of separation (DOS) from the Air Force. I knew that if I was assigned to fly a "heavy" aircraft after the war I would only have to fly that airplane until my time in the Air Force was up. That would be two years from the date I left Vietnam. My own career decision process really paid off for me. Right on cue I was selected for the KC-135 tanker training in California for an assignment to a squadron in Lockbourne, Ohio; this would start my two-year career flying a tanker aircraft. But a miracle happened just one month before I was to leave Vietnam: my assignment to the KC-135 squadron was canceled and I was reassigned to the Armament Development Test Center at Eglin Air Force Base in Florida. I spoke in detail about that assignment in the mission chapter of this book. My decision to put in a DOS paid off and that set me on a course to greater things in my Air Force career as one year after arriving in Florida I was offered an opportunity to withdraw may DOS and remain in the Air Force.

My next key career decision came when I was able to basically establish my next career move to the Air Defense Weapons Center (ADWC) at Holloman AFB in Alamogordo, New Mexico. While I was stationed at Eglin AFB, we had a recurring requirement to supply a couple of aircraft and pilots to support the test mission at the 6585th Test Wing at Holloman. The pilots would take their aircraft and stay for a one-month temporary assignment flying whatever test mission the 6585th Test Wing or the ADWC required. I knew this not only cost the government a lot of money but was a pain for the Test Wing at Eglin to send two aircraft and two pilots every month to support these test flights. I designed a plan and presented it to my operations officer, who I had previously worked with in the Flight Safety Office at Eglin. I would be willing to permanently move to Holloman and fly whatever mission needed to be supported and, in addition, I would be assigned to the ADWC as a quality assurance evaluator with the PQM-102 drone contract. It was a perfect solution to a problem for the Test Wing at Eglin, the Test Wing at Holloman, and ADWC detachment at Holloman. I was able to gain some very valuable career experience in the maintenance and contracting areas during this three-year assignment while still getting to fly many times a week. This career decision also allowed me to complete my master's degree in business administration and earn a promotion to major.

Through my career decision making process I was exposed to some key experiences that led to future promotions in the Air Force and my consulting career. I was getting ready to move from Holloman to my next assignment with an F-106 Delta Dart squadron in Tacoma, Washington. My wife was pregnant with our third child, and I had sent her ahead to Washington to find a home for us while I completed my three-month training in Florida for the F-106. Just before reporting to the F-106 training the Air Force pilot assignment office called to ask me if I wanted to change my assignment to an F-15 Eagle squadron in Virginia. Since I had already sent my family to Washington, I firmly declined the offer. While I would have loved to fly the F-15, I had a family to consider. That career decision really paid off for me as I was able to gain experience in the aircraft maintenance career field; I was also in a position to make a name for myself by leading the squadron conversion from the F-106 to the F-15. These two very important experiences really helped me in my future Air Force and consulting careers. Making key decisions about my own career certainly paid dividends for me and my family in the long run.

Arctic Weather Decision Making

One of the most intense decision-making times I faced was during the coldest weather ever recorded at Galena Airport during the month of January 1989. Hundreds of life and death decisions had to be made much like my combat time in Southeast Asia. These January low temperatures broke the all-time coldest day record that was set in 1961. To give you an idea of what we dealt with, the temperature range from 15 January to 31 January was a high of -36°F to a low of -70°F. The coldest period in January was 20 January through 27 January with the temperature range from -55°F to -70°F. We certainly had many issues with those temperatures, and I will describe in detail some of the actions we had to take to save Galena from the deep freeze.

The first indication of impending disaster came on 12 January during lunch when I got a call that the water pipes in the Galena Consolidated Club had burst and there was a great deal of water flowing out the front door and immediately freezing into a pool the size of a skating rink. The civil engineering folks were on the problem in less than thirty minutes and turned off the water to the building. The cleanup began in earnest, but it was still a huge mess. The weather that morning was -43°F, although that was not an unusual temperature for Galena in January. I consulted with my weather team, and they said that since all of Alaska was under a huge high-pressure area, the temperatures were forecast to be well below freezing for the next two to three weeks.

The temperatures continued to drop and on 16 January we hit -64°F and many problems followed. Calls poured in that the pipes in our communications building, the Combat Alert Cell, the junior enlisted dorm, the base operations building, and the air traffic control tower had frozen; many pipes were broken. On top of that we were getting calls that almost all of the vehicles that were kept outside would not start and or the wheel bearings were frozen solid. We had a problem on our hands, and I immediately called an emergency staff meeting with all of my officers and senior enlisted staff to come up with a plan to combat these extremely low temperatures. We had no preexisting checklists to deal with this freezing weather, so we were kind of "flying by the seats of our pants" to come up with a plan.

While we were convened in the staff meeting, I was notified that the dining hall, the transportation office, and the power plant office all had frozen pipes. Additionally, we were notified by the civilian air terminal that they also had frozen pipes and the Mark Air refueling station was frozen solid; there would be no more commercial air traffic into or out of Galena until further notice. I coordinated with the 21st Tactical Fighter Wing and the Alaska NORAD Region and recommended the F-15s be placed on Mandatory Scramble Order (MSO) status because the pipes in the pilots' living quarters at the CAC had frozen, effecting the alert pilots' ability to get the required crew rest before flying. Not long after, my operations officer called asking that I meet him on the runway for a problem he had detected.

My staff vehicle, a 1984 Chevy Suburban, barely started but I was able to meet the ops officer on the runway, and what I found was very scary. He told me to turn off the vehicle and just listen. We heard these loud popping sounds (like popcorn popping) coming from the runway and saw stones the size of quarters come flying out of the runway that had been paved just a few weeks before. The moisture in the pavement was freezing and the runway was literally coming apart. Large horizontal cracks appeared that were six to eight feet long, two to three inches wide, and five to eight inches deep. We were in "deep shit" at Galena. The temperature was -63°F. And that was just the start of an incredible sixteen days of trying to save Galena. As soon as I returned to the office, I notified the wing commander at Elmendorf that we had a major problem with the runway at Galena and that most of our buildings were freezing up and could soon be non-operational. The commander told me to get him a list of the things we needed to keep Galena open. I reconvened my staff and went around the room asking each functional leader what they needed to save Galena. Before leaving my command vehicle outside again I asked my first sergeant to find me a young, enlisted airman who would keep my vehicle running by driving it around base; this turned out to be another wise decision because, even if the engine was running, the wheel bearings froze in most of the vehicles left outside.

This issue became critical as we needed to decide what vehicles were absolutely mandatory for the current cold weather emergency and which ones could be "sacrificed." The list was long and included all fire department vehicles, four security police vehicles, numerous civil engineering vehicles, some aerospace ground equipment, the base water truck (which turned out to be a prophetic decision) and, of course, the commander's vehicle. Where we could, we assigned drivers to keep some of these critical vehicles moving and the rest we were able to place in a large hangar with a few portable heaters running nonstop. That seemed like a good plan until we got word that our vehicle refueling building and main refueling pumps on base were now frozen solid. We now had no way to refuel our running vehicles.

Friday, 20 January was another very cold day at -60°F. And we started off again with another cold-related crisis. The missile storage building was frozen, the headquarters building was frozen, and the metal handles on the headquarters front doors were so cold that they just broke off as we tried to open the building. We also learned a valuable lesson that day about very cold temperatures and the technology in the newer vehicles: if a post-1985 vehicle stalled and the maintenance experts carelessly closed the hood, all of the plastic components near the engine or the various drive belts would shatter. I gave new directions to the transportation folks to just leave the "dead" vehicles where they were unless they were causing a traffic issue.

Just when we didn't think anything could get any worse, I was informed that the diesel fuel that powered our generators was starting to gel at -54°F degrees, which meant that the fuel could not flow into the generators. This gelling process had caused a fuel spill at the power plant as the fuel was backing up in the transfer lines from the external fuel tanks to the internal storage tanks of the power plant. We were now down to just two of the four generators that were providing power to the base. I called an emergency meeting with the power plant operators and tried to get a solution to this very critical problem. Fortunately, one of the more seasoned power plant operators said he thought if we heated the diesel fuel to above -54°F degrees while it was pumped from the external storage tanks to the generators, we might be able to keep the fuel flowing. I called back to Elmendorf and requested that ten Herman Nelson 250,000 BTU/hour gas-powered portable heaters be sent immediately to Galena or we may have to abandon the base. The wing got right on the problem and dispatched a C-130 with the heaters. I had to temporarily open the runway for the C-130 to land, but we had our heaters in less than six hours to solve the freezing fuel problem. The civil engineer experts placed heaters near the output fuel line from the large external fuel tanks and near the intake fuel line to the generators, and the heated diesel fuel began to flow into the generators to save the day and the base.

Just as we were dealing with the power plant issue, we got a fire alarm notification from the alert hangar that all the fire alarms were activating at the same time. Fortunately, the water fire suppression system was frozen, so no water was sprayed on the two alert F-15s. I immediately ordered the F-15 pilots to evacuate the alert hangar and taxi the aircraft away from the building until we could determine the cause of the activation of the fire alarms. My civil engineer was convinced that the fire alarm was cold weather-related, and that his team would have it fixed in short order. We had the two F-15s keep their engines running until we could solve the problem. It was such an amazing site as the two aircraft started their engines and taxied out, they were both sending huge contrails, just like you see aircraft in the sky out of the back of the airplanes. The temperature during this crisis was -60°F.

After this fire drill was under control, I picked up my civil engineering officer to do a tour of the base and get an idea of the amount of damage we were sustaining from this cold weather. We stopped in front of an old barracks building that had been closed for the past year awaiting demolition. The civil engineer said, "You need to hear and see this incredible sight." As we entered the building, I could hear a very loud cracking sound followed a few minutes later by another loud cracking sound. It seemed that this building was freezing as we walked through it and the noise we were hearing was the abandoned toilets in each room exploding as the frozen water expanded and cracked the porcelain water tanks.

More freezing issues occurred as one of the two main boilers that provided heat to the base went down. We were without full-capacity heat for nearly forty minutes and that really exacerbated the freezing of additional buildings on Galena. I called for another emergency staff meeting and for the first time addressed the possibility that we would have to abandon Galena if we didn't get any relief from the cold. I asked each functional area leader to designate the emergency personnel that we would leave behind if the base needed to be abandoned. We identified fifty people who we would keep behind if we did have to evacuate. The personnel department made travel orders for the remaining 270 personnel who we would evacuate to Elmendorf if necessary. The temperature was now down to -70°F, a new low record for Galena. With a five-mile-per-hour northwest wind, the wind chill temperature was down to a "feels like" -88°F. The large hangar, which we called the Birchwood Hangar, sustained more damage as the main civil engineering supply room located inside was flooded from a broken pipe. More vehicles died with frozen transmissions and wheel bearings keeping the majority of the motor pool fleet grounded; the only road grader in the base inventory was destroyed while crews were trying to level one of the main roads by the power plant. The cold weather caused the twelve-inch-wide and two-inch-thick steel frame on the grader's blade to crack completely in half. Dealing with these many cold weather issues tested my leadership abilities to the fullest and allowed the base to be saved from the record-breaking arctic freeze.

Critical Decision Making

One very good example of decision making and the tough decisions a leader must make came from an incident while I was the base commander in Alaska. A unique aspect of doing major construction work at Galena was that we were completely dependent on the status of the Yukon River. We could only get some of the large construction equipment on site by river barge during the time the river was open and flowing, usually between May and September. This was the case for some extensive concrete work that had been programmed for Galena for the past two years. Over three years, five new functional buildings at Galena had been constructed, but purchasing and installing the concrete for sidewalks had been put off for budget reasons. The paths to and from these new buildings were a muddy quagmire for the airmen trying to get from one building to the next. So, when the concrete plant — which had to come by barge — arrived and docked at Galena, my civil engineering officer said that we needed to take advantage of river remaining open for the near future.

Once the river started to freeze the concrete plant would have to move back upriver before it got stuck until the next ice breakup. The concrete plant operators agreed; however, we had already used up the money allocated for the current year for the concrete work that was desperately needed at Galena. I consulted with my civil engineering officer and my resource management officer and they said we could divert some of the money allocated for other base projects to take advantage of the concrete plant during the open river situation.

The weather was cooperating, in spite of summer rain showers, and we were making great progress on the new sidewalks. Late on the third day of the concrete pour project I got an emergency call from the head of contracting at Elmendorf. He told me to immediately cease all concrete work until he and his staff could come out to Galena. Reluctantly, I ordered the concrete work to cease but to keep the concrete plant in port until this problem was all sorted out. The next day the head of contracting and three of his staff, all in business suits, arrived at Galena during a heavy summer rainstorm. I took them to my office and heard their pitch that I had broken federal law by spending money that was not allocated for that fiscal year. I told them I was aware of that but felt mission requirements justified moving money from other projects to take advantage of the concrete plant while it was able to remain in Galena. I got some very stern looks, so I told them to come with me and look at the concrete project.

Together we drove in my command vehicle, and I parked just across the street from where the concrete work that was being done. Remember that it was still raining, and the road and non-concrete paths were very muddy. I told the Elmendorf staff that the squadron airmen visited the post office, their dorms, and the dining hall on a daily basis and they had to walk through the mud. I encouraged each one of them to follow me as I walked from the post office to the dining hall. They were none too happy with the walk as they all had mud flowing over their shiny dress shoes and up their pant legs. As we got to the dining hall for our scheduled lunch, the conversation shifted to, "We didn't know the problem was this bad at Galena without these sidewalks." As lunch was finishing up and it was time to get them back to their aircraft for their return to Elmendorf, the head of contracting told me to continue to pour the concrete as long as the concrete plant remained docked in Galena. He also said that he would find the additional funds ($150,000) to pay for the over-budget concrete project. This turned out to be a huge win for the squadron as the airmen no longer had to trudge through the mud during their daily activities. As soon as the contracting experts departed Galena, I called the wing commander and told him I was not going to prison for the concrete work. Sometimes a leader must make the tough decisions and then be prepared to live with the consequences. This concrete incident had a positive ending, but I was ready to take responsibility for whatever may have transpired.

Decision Making Leadership Lessons

- Timely decision making is critical.
- Decide… do it now.
- Don't delay critical decisions.
- Take responsibility for your decisions.
- Don't wait on perfect information… it will never come.
- Change the decision if better information does become available.
- A decision delayed is no decision.

9

DELEGATION

Delegation is an important part of my leadership philosophy. An effective leader who is true to himself and the mission realizes that he just can't do everything expected of a leader. Delegation is simply assigning a specific task or group of tasks to a subordinate to assist in accomplishing the assigned mission. My experience both as a follower and a leader tell me that the most efficient organizations are those that have a healthy delegation process from the overall leader and some of the lower-level leaders as well. Delegation in my leadership model gives the leader four key tools: more time for strategic planning (long term), the ability to prioritize the tasks that need to be accomplished, empowered followers who are involved directly in the mission, and an expansion of the team concept — which I will cover in the last chapter of this book. The short vignettes in this chapter will explain the delegation process as I saw it in many organizations, both military and civilian.

Combat Delegation

During my year flying combat in Southeast Asia the concept of delegation became crystal clear to me. I was a trained and very proficient pilot in my early twenties and I quickly came to the realization that I could not do everything myself on every combat mission. While I was the on-scene commander in many situations there was no way that I could control or direct all aspects of every mission. There were many actions that I had to delegate to others in the hope that all of the pieces would come together for a successful outcome. I will cover this concept in more detail in a related chapter on teamwork.

When I speak of delegation in a combat environment, I am referring to the many tasks that others must perform perfectly in conjunction with my actions. Those actions ran the gamut from weather forecasting and radio discipline, to correct acquisition of the targets by the fighter aircraft and the flying ability of the rescue helicopter pilots. I just couldn't do all those tasks by myself, and the concept of delegation was really cemented in my mind and my future leadership opportunities. While it took me a couple of months to be comfortable with the delegation concept in combat, I began to realize that I had to trust the capabilities of all the players in the process if I expected a satisfactory outcome. This reliance on other parts of the mission allowed me the time I needed to concentrate on the important task at hand: supporting our ground troops in contact with the enemy.

Delegation in Flying

While my 750-plus hours of combat flying in Southeast Asia gave me great insight into the concept of delegation, I learned a valuable delegation lesson when I returned to the States and began flying a multi-crew aircraft. Coordinating crew actions while flying a multi-crew aircraft was a real challenge for me as I had flown by myself on nearly every combat mission in the war. Flying with another crew member involved a very high level of delegation for the pilot in command. The copilot or weapons systems operator (WSO) in the right or back seat had specific duties to perform for the safe operation of the aircraft. Some of those duties were delegated by the pilot in command during the preflight briefing. Part of the delegation process in flying is to identify and communicate some of the specific duties that will be delegated during flight, for example the process of ejecting and the exact terminology that would be used to initiate an ejection if that becomes necessary. In two-pilot aircraft, decisions need to be made about who talks on the radio and who calls out altitude, runway alignment, aircraft configuration, and air speed. Even though the pilot in command is ultimately responsible for the entire flight, the delegated duties to the other crew member are critical to a safe flight.

On one of the multi-crew aircraft that I flew we had an additional crew member, a flight engineer, who was a critical part of the flying operation. He would manipulate the throttles of the aircraft and set a specific speed requested by the pilot in command. That duty delegated to the flight engineer was a crucial part of safely taking off and landing the aircraft. I learned a great deal from this delegation of duties that I would apply in many leadership situations later in my Air Force and civilian career.

Day-to-Day Delegation

In my twelve different assignments in the Air Force, I experienced the entire spectrum of the delegation process from almost none to nearly complete delegation of various duties. In each of those organizations the middle ground delegation process worked the best. I worked for bosses where no matter what was delegated, the boss would almost always take over the project and put his own spin on the final product. Certainly, this did not give me the motivation to do a good job the first time as I knew the boss would always change the project. If only we knew what he wanted at the initial tasking we could have both been much more effective and saved a lot of time.

When I was the chief of a large maintenance branch in a fighter aircraft squadron, I learned immediately that I had to delegate many of the duties necessary for the branch to be successful. I relied on the professional knowledge and expertise of three to four senior enlisted members within a particular function of the branch These functions that I was responsible for included back shop avionics repair, the storage and maintenance of nuclear weapons, the life support shop, the parachute shop, and the F-106 simulator operations and maintenance. There was just no way that I could be the expert in each of these areas. My enlisted leaders came through and provided the squadron with the most professional support possible. This assignment really laid the foundation for the delegation concept that I used later in my career when I was twice selected as a squadron commander.

Commander Delegation

I was fortunate to have two squadron commander assignments in my twenty-six-year Air Force career. The concept of delegation was never more important than during those two assignments. A commander is the one person who has the ultimate authority and responsibility for every action in an organization. As I soon found out, there is no way that a commander can do everything in an organization and must rely on the delegated duties of many people. I will outline the delegation process in the two commander positions, first as a commander of a tactical flying squadron in Alaska and second as a base commander in Alaska. Both assignments had challenges but because I understood the value of delegation, I was able to effectively lead these two organizations through some difficult and trying situations.

Tactical Flying Squadron

I was fortunate to lead the 5021st Tactical Operations Squadron for almost twenty months. The Spyders, as the squadron as known, were tasked to provide all tactical air-to-air combat flying support to the two F-15 squadrons assigned in Alaska. Our specific mission was to simulate Soviet air-to-air combat tactics and really challenge the F-15 pilots in almost daily dissimilar air combat training (DCAT). It was a real challenge flying this mission in very old aircraft flown by very young Air Force pilots. As an experienced pilot my initial inclination was to be intimately involved in the day-to-day operations of the squadron. I soon learned that there were many other commander duties that I needed to tend to that would certainly affect my ability to also run the operations of the squadron.

As I had seen in other flying squadrons I had been assigned to earlier in my career, I turned to my operations officer and had a very detailed discussion about the role I would play in the squadron. I explained that he, the operations officer, would be in charge of all flying operations. I intended to stay out of his way, but I would also give him feedback on my impression of how things were going in the flying aspect of the squadron. I explained that I would take care of all other major activities in the squadron, and he would be in charge of all components of the flying operations. I was very blessed to have a talented operations officer and all flying activity went very smoothly while I was in command of the squadron. I would sit down with this officer once a week to review flying activities and would offer, only when necessary, ideas or comments on the operations of the squadron.

I did have another technique that would "save" me from some of the rather mundane duties I needed to accomplish beyond the flying mission of the squadron. Whenever I had a boring financial planning meeting scheduled with the wing commander or his staff that coincided with a critical project I was working on, I would have the operations officer call me on my brick — a large hand-held radio that key leadership carried each day — and say I needed to get back to the squadron. The brick calls saved me from many boring meetings.

In any organization there are many issues a leader must deal with. My flying squadron consisted of thirty officer pilots and fifteen enlisted support personnel; while this was not a big squadron the normal personnel issues were always present. In order to handle these issues I designated a highly talented master sergeant as my first sergeant. The most important role of a first sergeant is taking care of the airmen in the squadron. To be sure, taking care of airmen is the responsibility of everyone in leadership and supervisory positions in the Air Force, but for first sergeants, taking care of airmen is their primary responsibility. The first sergeant is the eyes and ears of a squadron and serves as his or her commander's critical link for all matters concerning enlisted members. First sergeants are responsible for providing sound advice to the commander on a wide range of topics including the health, *esprit de corps*, discipline, mentoring, well-being, career progression, recognition, and professional development of all enlisted members. First sergeants must respond to the needs of airmen twenty-four hours a day, seven days a week. Another important aspect of the first sergeant role is the relationship with the commander. First sergeants and commanders are not friends; their relationship is far more important than that. The bond between the two is based on mutual trust and respect. To be effective, the first sergeant and the commander must work together as a team through honest dialogue and the right to disagree when necessary. So, with all these required responsibilities a first sergeant is expected to do it was a very easy task for me to delegate the hands-on care of our airmen to the first sergeant. This delegation by no means took me out of the equation in supporting our airmen; my first sergeant and I had two to three meetings a week dealing with one of the many personnel issues normally found in any squadron in the Air Force. While the first sergeant had the first direct contact with the airmen in any situation, I always had the last say in the process as the commander.

This collaboration and coordination with the first sergeant worked very well and reinforced my leadership concept that delegation is critical in any organization. This coordination model became even more critical in my assignment as a base commander just six months following the flying commander assignment. As a base commander I was responsible for every aspect of the base from the air defense alert mission and the maintenance support for the alert aircraft and munitions, to housing, food services, civil engineering, air traffic control, communications, and many more important tasks. So needless to say, the delegation process was alive and well as I served as base commander in Galena, Alaska.

Deployed Delegation

Out of necessity my first day in command at Galena Airport required me to make a huge leap of delegation. A large portion of our air defense alert personnel were deployed to Eielson AFB in Fairbanks, Alaska because our only runway at Galena was being repaired and resurfaced. Fortunately, I had experience in delegating the operational mission of my squadron in my previous assignment as commander of the Aggressor Squadron, the Spyders, in Alaska. But it was still a stretch for me to give complete control of our air defense alert mission to my operations officer, a major, who was nearly 300 miles from my location. On the second day after I assumed command I hopped on an airplane and flew to Eielson AFB to meet with my operations officer and the fifty-six deployed squadron members. Even though this group was not scheduled to return to Galena for another six weeks I felt it was imperative to meet with the deployed staff and give them my command philosophy.

The operations officer set a up a special meeting, a Commander's Call, for the deployed personnel, and I covered in great detail what my expectations were for the remainder of the time they were deployed. I had a very receptive audience, and I believe they were pleased that

I came to share my leadership philosophy with them. My operations officer and I had a very detailed follow-on meeting as I described my delegation philosophy with him, which included empowering him with the authority to run operations the way he saw fit, to take whatever he thought was the correct action, and to contact me for any guidance if he had any questions. I also reiterated to him that I would back him 100 percent. We had absolutely no issues with the deployed operation, and during those six weeks we intercepted six Soviet bombers testing our response to their flights near our airspace.

Crisis Delegation

One of the many challenges that I faced as a commander of a remote forward-operating location was the extreme weather that we dealt with in January 1989. We faced a record-breaking -70°F for nearly two weeks. Many of the base functions were frozen solid, and I was contemplating evacuating the 320 members of the squadron back to Anchorage to save lives and equipment. During this extreme weather condition, the City of Galena requested emergency base support as the local residents were also freezing to death. Their water supply was completely frozen, the city was running out of firewood, and food deliveries to the city were at a standstill. Some of the locals even resorted to killing and eating their sled dogs just to survive. When the request for help came from the Galena mayor, I was in a very difficult situation trying to deal with both the crisis on base and the crisis in the city. After a quick trip to the city to estimate their emergency needs, I knew I needed to concentrate all my leadership efforts on saving the base. I knew that I needed to delegate this city problem to one of my staff members.

Fortunately for me, my base chaplain and first sergeant immediately stepped forward and volunteered to assist the city in any way they could. They did an all-base request for excess firewood and anything that could burn including excess oak furniture from a condemned base barracks and hundreds of wooden shipping pallets. We delivered more than seven truckloads of wood to the city. The chaplain and first sergeant coordinated to transport our emergency water truck into town and deliver 15,000 gallons of water to the city. They also rounded up nearly 145 cases of field rations (Meal, Ready to Eat or MRE) and collected hundreds of pounds of excess cold weather clothing donated by the members of the base. This effort again reinforced my belief in the amazing things delegation can do for a leader. Their efforts certainly allowed me to dedicate all of my time to saving the base from the extreme cold weather. After almost two weeks of near-record cold temperatures the weather finally broke to a balmy -40°F, and we tried to get the base and the city back to normal.

Volunteer Delegation

In my role as the overall coordinator and leader for a Christian lay ministry program I quickly realized that I could not do all the tasks necessary to have a successful ministry. I was still working full time as an aerospace industry consultant and was on the road too many days a month to provide hands-on leadership for this ministry. There were many facets of the program, from recruiting new ministers, interviewing the people receiving care, and assigning ministers for those seeking care, to training the new volunteers, presenting a dynamic and relevant continuing education program twice a month, supervising the small groups, and maintaining a support library with pertinent material and ministry forms. I knew I needed others to assist in this endeavor. Fortunately, the basic outline and structure of the ministry program included recommendations about key personnel positions. I immediately selected four key volunteer leaders and assigned them the specific tasks listed above. The volunteers were sensational, and each person on the team took on the challenge of doing all the actions necessary to have a successful ministry. This formula has worked very well over the years as I just completed my twenty-fifth year leading this vital ministry program.

This ministry program reinforced an important aspect of my leadership model: if you find the right people and give them all the resources they need for a task, you will have very good results. Over the twenty-five years of leading this program, I have trained more than 175 ministers and of those I have been able to delegate management tasks by mentoring sixteen as functional leaders and small group supervisors.

Delegation Leadership Lessons

- Leaders can't do everything.
- Delegate to your best people.
- Prioritize what to delegate.
- Completely define the delegation.
- Trust the delegation process.
- Delegation enhances effectiveness.

10

TEAMWORK

In my leadership model each of the previous nine leadership traits that I described are critical to having a successful team. The concept of teamwork is so important in my leadership model that this idea really enhances all the actions a leader can take to have a successful organization. There is not really a single definition of a team, but generally the concept of a team involves getting people with different backgrounds and skills together to achieve a common goal or mission You need competent people on the team. The team must be focused on the mission or they are wasting time. You need to have the right people on the team. The team must communicate between the members and the overall leader to be successful. The team must follow the established standards and have the integrity to enforce those standards. The team must be loyal to each member and the overall leader. The team leader must have a level of effective decision making to enhance the overall leadership and be able to delegate tasks when they are needed to accomplish the assigned mission.

An important element to building successful teams is the leader's communication; the team needs to clearly know exactly what it is supposed to be doing. An old colonel I worked for in Florida gave me a very precise formula to use when setting up a team. I used this formula many times in my Air Force and civilian careers:

- The leader must ensure that each team member understands the overall goal of the team.
- The leader must establish clear deadlines.
- The leader must define the boundaries of the team.
- The leader must describe how each role of the team contributes to the overall goal.
- The leader must provide the desired knowledge and resources to complete the goal.
- The leader must ensure the team communicates up and down the chain of command.

I have seen this teamwork formula work so many times in many different leadership environments that I have incorporated teamwork as an overarching tenet of my entire leadership philosophy. The following stories define how teamwork played an important role in my career.

Combat Teamwork

As I mentioned early in this book one of my missions in my Air Force career was as a forward air controller during the war in Southeast Asia. Teamwork was so critical in every combat mission I flew that I thought the best way to describe the teamwork concept in combat was to review a mission from 22 September 1971. This mission was in support of an attempted prisoner of war (POW) rescue near the prison outside the Cambodian town of Kratie on the Mekong River. As I have described before, Kratie was a hot spot for enemy activity and a major staging location for NVA/VC supplies. There was a well-developed, concrete-wall prison located roughly four kilometers to the east of Kratie. My fellow FACs and I routinely flew over the prison and did visual reconnaissance every time we flew near or over Kratie. We also flew many High Low missions near Kratie. On one such High Low mission our low bird was able to take a photo of someone in the center courtyard of the Kratie prison waving what looked like a white tee shirt at the FAC. After the photo was developed it was determined that, in fact, there was someone in the center courtyard of the prison waving something white at the aircraft. This information was passed along to the Special Operations Group headquarters for action. It was well known that about thirteen U.S. troops were missing and potentially held captive by the NVA/VC during the U.S. invasion of Cambodia in May of 1970. The thought was that perhaps there could be an American prisoner still confined in the Kratie prison. The actual organization that conducted all suspected POW rescue missions was the Joint Personnel Recovery Center (JPRC). While officially part of SOG, on paper, the JPRC acted independently from the SOG for the most part but exclusively used the SOG teams for the POW rescue missions. These POW rescue missions had the code name Bright Light.

The FACs were ordered not to fly near the prison while a rescue plan was being put together by the SOG headquarters. We did not want to give any indication that we were interested in the prison. About four days later we got word that a rescue mission would take place during the night hours of 22 September. The mission involved a team of three U.S. Special Forces members, two Montagnard Special Forces, and one Vietnamese Special Forces soldier who would complete a HALO (high-altitude low opening) parachute drop near the prison and then attempt a rescue if there were American prisoners there. In a HALO jump the parachute team exits the back of C-130 Hercules aircraft at 13,000 to 18,000 feet and delays opening their parachutes until they are much lower in altitude. Each jumper stabilizes his fall and maneuvers (flies) for about one minute, opens the parachute at around 2,500 feet, and then descends for another two and a half minutes, flying the parachute to the designated drop zone. The concept was if the drop aircraft flew high enough the enemy on the ground would not know an activity was taking place in their area.

My part of this HALO mission that night was to be the radio relay for the dropped team so they could communicate through me to SOG headquarters. The weather, unfortunately, was less than perfect but I did get a call that the C-130E from the USAF 20th Special Operations Squadron had taken off from Saigon and the drop mission was on. The planned drop was to be from 13,000 feet, and I would be orbiting well to the south at 3,500 feet. My take off time was 0100 from the very dark Quan Loi runway. The drop was scheduled to occur at 0200, and I was to attempt radio contact with the team as soon as they all landed and had regrouped as a team. The designated landing zone was about 750 yards to the southeast of the prison. The plan was to keep in radio contact with the team throughout the night and then coordinate an extraction from a pre-selected LZ to the southeast of the prison after the mission there was complete. My schedule was to fly until 0600 and then be replaced by another FAC, Pretzel 03. I never did see the C-130 because of the weather in the area but I did get a call that the drop had occurred at 0207 and that SOG was just waiting for radio contact. Still not wanting to give their position away I stayed as far from the prison as I thought I could and still be in radio contact range with the team.

Finally, at 0310 I got a call from the assistant team leader known as One One on the ground. He said the team leader, known as One Zero, had been badly injured during the jump and he was still attempting to locate two of his other team members. Apparently, they had become separated during the jump. I advised the SOG folks of the current status and was told to stay on scene until the ground team leader found all his team and decided on a course of action. Finally, at around 0400 I got a call that the One Zero who said one of the allied SF troops were unable to continue due to injuries sustained in the HALO drop. After I relayed this information back to SOG headquarters I was immediately informed that the mission was an abort and to move the team to the primary extract landing zone.

The SOG headquarters was already in the process of putting the extract plan together and wanted to get the team out at first light. I got a visual contact with the team via a red strobe light and informed the team that they were approximately one kilometer from the briefed primary LZ. I informed One One they should take a southeasterly heading to get to the LZ. After further discussion with One One he informed me that the team could get to the LZ in about forty minutes as there were no enemy forces trailing them at the time and they needed to assist the injured team members. With a final radio check and strobe light contact I informed One One that the LZ was approximately one hundred meters on a heading of 135 degrees. The extract package consisted of two H-34 Kingbees and two Cobra gunship helicopters. Since the distance from Quan Loi to the LZ was only about fifteen miles I expected the helicopters to arrive at the LZ just before dawn. I was orbiting at around 2,500 feet at the time and got a good visual on the helicopters. I marked the LZ for the lead helicopter and Kingbee

27 said he had a visual on the LZ. The extract went off without a shot being fired at the rescue team or any shots from the attack helicopters. While the mission was a failure, we did get the rescue team out of harm's way; I don't believe this mission on the Kratie prison was ever completed, at least not during my time in Southeast Asia. Once again, however, this combat experience confirmed another key element of my leadership model… teamwork is an effective way to support whatever mission or task you are assigned.

Teamwork Saves the Day

One of the most challenging leadership tasks I faced was in May 1989 while I was the base commander of Galena Airport. Galena Airport sat right on the banks of the Yukon River. Each Spring following the harsh winter in Alaska, the frozen Yukon River would break up in massive ice floes and in certain years those floes caused very destructive flooding on its way to the Bering Sea. The break-up in May 1989 followed the record-breaking cold weather Galena faced in January. During that January we had a stretch that lasted for thirteen straight days when the daily temperatures hovered between -50°F and -70°F. The ice on the river was more than three feet thick in places.

Our Air Force mission at the Galena Airport was to have two F-15 fighter aircraft on 24/7 air defense alert to counter any Soviet air threat to the U.S. or Canada. There were usually about 320 to 350 military and civilian personnel assigned to Galena to support the air defense mission. We had a detailed checklist in place to help us deal with any potential flooding issues that may come from the annual river breakup. As the flooding on the Yukon progressed, we rapidly approached the flood level requiring evacuation for the majority of the personnel from Galena to Elmendorf AFB in Anchorage.

As we continued with the process of evacuation, I held an emergency staff meeting with the fifty-one personnel designated to stay behind at Galena and try to save the base from the flood waters. I explained that all of us who stayed behind had specific duties to accomplish and that there would be times when we would all have to join in and do whatever it took to save the base. What a tremendous response I received from the group we called the Yukon River Rats. There were so many tasks to perform and the teamwork shown by these fifty-one people was spectacular. We had administration personnel become security police members, we had cooks working with the fueling folks trying to keep the floodwaters out of the main pumping station, we had security personnel lend a hand in the chow hall, we had electricians doing plumbing work, and we had services personnel filling sandbags. The list goes on and on… and I did not hear a single complaint. These dedicated Yukon River Rats were able to keep up the support of the base for nearly ten days until the flood waters slowly began to recede, and things got back to normal once the 270 deployed personnel returned to Galena. I was so impressed with the teamwork of my squadron that I wrote an editorial for the base newspaper at Elmendorf AFB praising the amazing teamwork I saw during this difficult time. That editorial is below:

"As I reflect on my last 10 months as Commander of Galena Airport, the time frame from 13 January to 25 May stands out as the most intriguing, exciting, rewarding and memorable. It was during this time I witnessed the most graphic displays of teamwork ever demonstrated during a peacetime, real world period. It seemed like the men and women of the 5072nd Combat Support Squadron (CSS) thrived on the challenges they encountered; the more they were challenged, the better they performed."

"The period began when somebody turned the thermostat off. Temperatures plunged to an average of minus 60 degrees for 18 straight days, peaking at minus 70 degrees. Pipes froze or burst, vehicles had to be moved inside to start, two out of three power generators went down and the third was on the way down because of potential overload. Our dinning facility was under renovation and food had to be prepared at one location then transported to another for serving. Portable heaters had to be used to prevent the loss of our last generator and our heating fuel supply had reached a critical low point."

"Through it all we survived because of teamwork. We survived and the intercept mission of Galena remained on alert status the entire time. Our plumbers fixed pipe after pipe, the fuels people coordinated with Elmendorf to get a 700,000-gallon emergency air delivery of fuel, then off loaded it in minus 60-degree weather. While all of this was happening, the men and women of the 5072nd CSS made time to collect firewood and clothing for several elderly families in the local community. Each person in the squadron had to do their part for things to work out. It was a fine-tuned team that brought us through the coldest winter in Galena history."

"Shortly after the cold spell came the breakup of the ice on the frozen Yukon River. The mighty Yukon River with currents in excess of seven knots has the potential to devastate everything in its path when not confined to its natural path. We developed, reviewed and finally practiced our plan for an emergency flood evacuation. We were planning another exercise with our flood evacuation plan when Alaska State officials notified us the Yukon had risen to 140 feet above normal river level and was expected to top out at 154 feet within hours. Colonel H.S. Storer, 21st Tactical Fighter Wing Commander gave the word to "evacuate your people. Immediately base operations coordinated the launch of 6 C-130 Hercules aircraft from Elmendorf."

"The 51 people left behind at Galena began flood mitigation preparations. Complete activities were dismantled, computers from the first floor were relocated in offices and hallways upstairs; unneeded vehicles were moved to higher ground and boats were moved to key locations for emergency, last minute getaways. The Galena Operations Center began 24-hour operations. Fuels people became cops, food service people became projectionists, and civil engineers became administrators. While monitoring the river's activities, coordination with Elmendorf revealed the support to our evacuated members to be flawless. With very little advanced notice, Elmendorf provided aircraft and ground transportation for an unexpected 270 people. A section of Elmendorf's family housing area was instantly converted into temporary quarters."

"After nine days at Elmendorf the "all clear" signal was given and the men and women of the 5072nd CSS began the ordeal of returning to Galena and putting their lives and their mission back in order. But the 5072nd people were returning only seven days before the Alaskan Air Command Unit Effectiveness Inspection (UEI) team was scheduled to inspect our squadron. The final week before the UEI we had to unpack complete functions, reassemble flight line equipment and activities, relocate and reconnect computers. On top of this we had to complete work on the UEI work center, organize billeting and transportation for the inspection team and refine everyday procedures to meet the very in-depth review Galena would receive. Within five days after returning from Elmendorf, things fell into place. Equipment was back online, and final tests of radar and flight line equipment were complete. Meanwhile our alert aircraft had returned and the fine tuning of everyday operations within five days of our return from Elmendorf was complete.

"The UEI Team stayed for four days. The result of the teamwork during the UEI resulted in an overall squadron rating of EXCELLENT. These examples serve as a constant reminder to me of the incredible results that are possible when a unit works together as a team. This group of professionals banded together against time, the elements and the odds to prove that anything is possible when you do it as a team.

"While I am, extremely proud of the men and women of the 5072nd CSS I would also like to complement all the people of Elmendorf for their outstanding support and teamwork. It was their combined teamwork which enabled the airmen of Galena to overcome some incredible challenges and succeed as we have. Teamwork then is a very important ingredient in facing and overcoming tremendous challenges. The people at Galena know it works and so do the people at Elmendorf. We know how to do the Top Cover for North America mission."

Consulting Teamwork

Toward the end of my consulting career, between 2005 and 2010, the government changed their contracting methodology a great deal and came up with the task order contract concept. In this new task order process, the government consolidated many different support functions into one major contract with the hope that multiple contracting companies would join forces and perform the contract work whenever the government decided it was time to do a particular function. For the most part the concept was sound, from a contracting point of view. The government would have a standing contract in place before a specific service or product requirement was even known. The task order contract required the selected contractor to provide services or products as ordered by the government from time to time. For the companies that tried to win the contracts it was a pain in the ass and cost millions just in bid preparation fees. But these were the new contracting rules, and to win a contract the companies had to play the government contracting game.

The principal idea for the task order concept was for four or five prime companies, like Boeing, Lockheed Martin, IBM, or TRW to be the lead company putting together the bid for the contract. The prime companies would add multiple other smaller companies to the proposal to cover the need for a specific expertise. This task order contract concept certainly changed my approach to finding government work for my clients. Prior to this new contracting concept, I would be "chasing" multiple different parts of many diverse contracts, anticipating that my clients could do the specific work required of a particular contract. After the new concept was put in place, my main function was to find those pieces of work that would be part of a larger contract that the prime contractor could not do or did not want to do. I had to be very wise in meeting with the potential prime contractors that would bid on the task order contract with the hopes that my client's particular expertise would be a major selling point on the team we joined to help win the contract.

So, this I where the teamwork concept came into play. As the various teams worked to prepare a bid for the government proposal, each part of the potential tasks of the government proposal needed to be covered in the bid the teams submitted. There were many hours of sometimes intense negotiation among the team players on a contract proposal as each company tried to carve out a particular part of the proposal. The trust among the team players was a critical part of the process. I had to make sure that the prime contractor would in fact assign the smaller team members a piece of the contract once it was requested by the government and that my clients would perform the work satisfactorily. I soon learned to trust only a few of the prime companies as their track record on some of the initial task order contracts did not favor the small companies and very few tasks were passed along. As this process moved forward and covered many of the various contracts that were coming out of NORAD and U.S. Space Command, I found that the companies that worked together as a team on the task order contracts became very successful and the government noticed. While cost was always a factor, there were times when the government awarded a task order contract based on the collaboration among the various companies. So again, the concept of teamwork was alive and well and is a very important part of my leadership model.

Volunteer Teamwork

When I first was involved in the Christian lay ministry program at our church, five of the six ministry leaders were all working full time at their various professions. Having the time to build and maintain this ministry took some very delicate time management. However, in spite of the various time constraints of the leaders, we were able to build a team framework that was more than adequate to make this ministry a success. We got all the leaders together to formulate and approve a team model whereby each one of the leaders would be assigned a specific task to carry out. The functional leader roles included me as the overall coordinator along with the referrals coordinator, the supervision coordinator, the training coordinator, the continuing education coordinator, and the library and support coordinator. Each leader had specific defined tasks to carry out each week. As the overall coordinator my job was to make sure I stayed out of the way of the other leaders and to remove any obstacles they encountered while doing their tasks. Team coordination was an important factor in our success. We were all striving for the same outcome.

After the first six months of this team action, we could see that our program was going to be successful. This team concept was built on a couple of previous leadership traits I have already described, from doing the assigned ministry mission and maintaining the established standards of the program, to picking the right people for the various leadership roles and having the integrity to trust and depend on the actions of each team member, to facilitating communication among the team members. In a way, this ministry team was a true picture of how a leadership model can contribute to the success of any endeavor.

Teamwork Leadership Lessons

- Teamwork is effective every time.
- Define exactly what the team's goal is.
- Define specific milestones for the team to meet.
- Encourage teamwork.
- Reward teamwork.
- Teamwork reduces the overall workload.
- Coordinate the team action.
- Communicate between the team members.

While teamwork is the final tenet of my leadership model it is by no means the only tenet that a leader should use to be successful. The organization or situation a leader finds him or herself in will dictate just how much of each leadership tenet they can implement. My fifty-plus years of leadership experience certainly showed me that you must be flexible in some of your leadership skills and be ready to modify some leadership approaches based on the mission, the people, and the situation. The situations I used to describe my leadership tenets were based on my many years of hands-on training, learning both the good and bad side of a leadership situation. I hope that some of this information in this book will help you in any leadership situation you may encounter.

ABOUT THE AUTHOR

Tom Petitmermet has over 50 years of leadership experience, including 26 years in the U.S. Air Force, where he spent time flying combat missions in Southeast Asia and held many important command and staff leadership roles. After his retirement, he applied these skills in the civilian world as a defense contractor that provided support to various technical programs, and also applied his leadership knowledge as a key volunteer for other organizations. In his third book, Tom shares the ten tenets of the leadership model that he developed as he learned firsthand how to lead others in the real world.

ALSO BY TOM PETITMERMET

Find Tom's books on Tactical16.com
or Amazon.com!

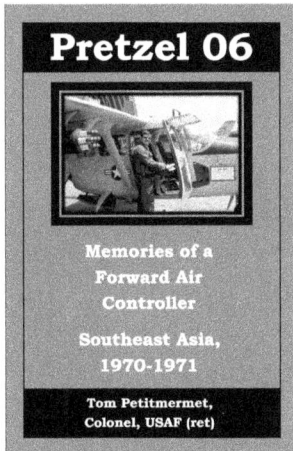

Pretzel 06: Memories of
a Forward Air Controller
Southeast Asia 1970-1971

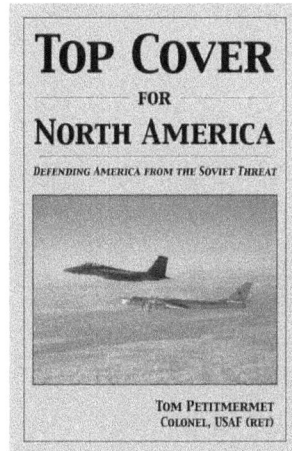

Top Cover for North America:
Protecting America from
the Soviet Threat

ABOUT THE PUBLISHER

TACTICAL 16

Tactical 16 Publishing is an unconventional publisher that understands the therapeutic value inherent in writing. We help veterans, first responders, and their families and friends to tell their stories using their words.

We are on a mission to capture the history of America's heroes: stories about sacrifices during chaos, humor amid tragedy, and victories learned from experiences not readily recreated — real stories from real people.

Tactical 16 has published books in leadership, business, fiction, and children's genres. We produce all types of works, from self-help to memoirs that preserve unique stories not yet told.

You don't have to be a polished author to join our ranks. If you can write with passion and be unapologetic, we want to talk. Go to Tactical16.com to contact us and to learn more.

TACTICAL 16
PUBLISHING